CHEMICAL ENGINEERING METHODS AND TECHNOLOGY

THE EFFECTS AND PERFORMANCE ANALYSIS OF NON-LINEAR PHASE NOISE IN ALL OPTICAL OFDM SYSTEMS

CHEMICAL ENGINEERING METHODS AND TECHNOLOGY

Additional books in this series can be found on Nova's website under the Series tab.

Additional e-books in this series can be found on Nova's website under the eBooks tab.

CHEMICAL ENGINEERING METHODS AND TECHNOLOGY

THE EFFECTS AND PERFORMANCE ANALYSIS OF NON-LINEAR PHASE NOISE IN ALL OPTICAL OFDM SYSTEMS

IRAJ SADEGH AMIRI

AMIN KHODAEI

AND

VOLKER J. SORGER

novinka
New York

NOTICE TO THE READER

Library of Congress Cataloging-in-Publication Data

ISBN: 978-1-53613-145-1

Published by Nova Science Publishers, Inc. † New York

CONTENTS

PREFACE

Due to the limitation of the electrical OFDM signal and electrical Fast Fourier Transform (FFT), all-optical OFDMs have recently received much attention. Accordingly, this research study was conducted to investigate the effect of phase noise in the performance of an all-optical OFDM transmission system with 4-point FFT single mode fiber (SMF) links by considering the effects of fiber length, input laser power and a number of channels. In all optical systems, the transmitter side consists of a comb power generator, wavelength selected switch and an optical QAM generator. A comb power generator generates channels with a frequency separation of $\Delta f=25$ GHz. Subsequently, a Wavelength Selected Switch (WSS) was used to split subcarriers and then the subcarriers were modulated individually with Optical QAM modulators. As the results show, a higher number of channels led more phase noise in terms of XPM and FWM nonlinearities, and signal power was the main factor in nonlinear fiber optics. As a consequence, there is more phase noise distortion at a higher signal power for a higher number of channels rather than the lower number of channels.

Chapter 1

INTRODUCTION
TO ALL-OPTICAL OFDM PROS AND CONS

ABSTRACT

Orthogonal frequency division multiplexing (OFDM) could be a promising technology in wireless and optical transmission system's space to hold high-data rates in optical communications. The orthogonality property of OFDM gives rise to scale high data rate digital signal into multiple lower speed subcarriers. Moreover, optical OFDM (OOFDM) causes many advantages in radio-over-fiber transmission systems. All-optical OFDM have received much attention recently to overcome the existing limitations in optical communication systems. The main issue in all-optical OFDM system is that the system suffers from phase noise effects and also being more sensitive to carrier frequency offset (CFO). In fiber-optic communication systems, the weaknesses of OFDM might result in higher sensitivity to fiber nonlinearities. In an all-optical OFDM transmission system, fiber nonlinearities affect the performance of the system significantly. The effects of fiber length, input laser power and number of channels is investigated in this book.

Keywords: all-optical OFDM systems, phase noise effects, fiber nonlinearities

1.1. ALL-OPTICAL OFDM ADVANTAGES
AND DRAWBACKS

Orthogonal frequency division multiplexing (OFDM) is a promising technology for high-speed wireless systems such as WiMax, third generation (3G) and fourth generation of cellular wireless network [1-8] as a Multi-Carrier Modulation (MCM) due to being high spectral efficiency and simple solution to transfer dispersive channel to multiple flat channels [9-12]. The orthogonality property of OFDM gives rise to scale high data rate digital signal into multiple lower speed subcarriers so that at the receiver, the OFDM signals demodulate with low complexed components and to reduce Inter-Symbol Interference (ISI), [13-15]. Despite the intriguing ability of OFDM, it has been recently considered for optical transmission systems [16, 17] due to many factors such as OFDM is much more resilient to dispersion compared to conventional Time-Division Multiplexing (TDM), it also conveys high-data rate or data-stream on a large number of subcarriers [18-21] and being more spectral efficient in comparison to wavelength division multiplexing (WDM) using orthogonality property of subcarriers [22-25]. Moreover, Optical OFDM (OOFDM) engenders many advantages such as immunity to Chromatic Dispersion (CD) and Polarization Mode Dispersion (PMD) in radio-over-fiber transmission systems [26-33].

Due to the fact that electrical OFDM signal is limited by two major components including optical modulator and photodiode in transmitter and receiver respectively; and electrical Fast Fourier Transform (FFT), all-optical OFDM have received much attention recently to overcome the above-mentioned limitation [34-38]. In general, Optical OFDM systems utilize electronics process in forward and inverse FFT/IFFT module, which requires high-speed digital signal processing (DSP), an analog-to-digital converter (ADC) and a digital-to-analog converter (DAC). Hence, electronics processes restrict Optical OFDM symbol modulation speed. On the other hand, Optical FFT typically is used at the highest speed beyond that state of art electrical FFT [39]. This is because of optical sampling window sizes using electro-absorption modulators (EAM) can be significantly shorter than its electronics counterpart [39]. Additionally, optical FFT is more efficient compared to its electronics counterpart in terms of power consumption because it uses passive components which excluding tuning circuitry and time gating.

The main issue in all-optical OFDM system is that the system suffers from Phase noise effects and also being more sensitive to Carrier Frequency Offset (CFO) [40] which creates Phase Rotate Term (PRT) on each subcarrier and result in inter-carrier interference (ICI) due to destroying orthogonality of subcarriers [41]. In fiber-optic communication systems, abovementioned weaknesses of OFDM might result in higher sensitivity to fiber nonlinearities such as self-phase modulation (SPM), cross-phase modulation (XPM) and four-wave mixing (FWM) [42-46]. Therefore, precise calculation of induced nonlinearity, which is produced by fiber dispersion, is crucial to cope with these side effects of OFDM.

1.2. INVESTIGATION OF FIBER NONLINEARITIES IN ALL-OPTICAL OFDM

In an all-optical OFDM transmission system, fiber nonlinearities such as self-phase modulation (SPM), cross-phase modulation (XPM) and four-wave mixing (FWM) affects the performance of the system [47-49]. Therefore, precise calculation of the nonlinearity produced by the fiber dispersion is crucial to cope with the side effects of OFDM. The main objective of this book is to investigate the effect of phase noise in performance on all-optical OFDM transmission system with 4-point FFT single mode fiber (SMF) links. The effects of fiber length, input laser power and number of channels is investigated in this research works.

This book is organized as follows; Chapter 2 presents a literature review on OFDM and all-optical OFDM. In chapter 3 System Configuration of Coupler based all Optical OFDM Transmission systems is presented. A detailed description of the all-optical OFDM system including transmitter and receiver components more details on comb power generator and optical QAM generator plus the characteristics of fiber link is explained in this part. In chapter 4 the effects of Kerr nonlinearity in optical fiber is presented. Analysis of the effect of SPM, XPM, and FWM on all-optical OFDM transmission system is included in this section. In this chapter, a theoretical analysis is presented starting with Maxwell equations to model the pulse propagation of the OFDM signal passing through anisotropic fiber. Numerical simulation results of all-optical OFDM, which are obtained from MATLAB and VPI Transmission software, are presented. In this chapter, the effect of the phase noise on the 4 Quadrature amplitude modulation (QAM) and 16 QAM is described. The effects of fiber length, input laser power and a number of channels are investigated at the end of Chapter 4.

REFERENCES

[1]   B. J. Schmidt, A. J. Lowery & J. Armstrong, (2008) "Experimental demonstrations of electronic dispersion compensation for long-haul transmission using direct-detection optical OFDM," *Journal of Lightwave Technology*, 26(1), 196-203.

[2]   Y. Benlachtar, G. Gavioli, V. Mikhailov & R. I. Killey, (2008) "Experimental investigation of SPM in long-haul direct-detection OFDM systems," *Optics express*, 16(20), 15477-15482.

[3] J. Armstrong, (2009) "OFDM for optical communications," *Journal of lightwave technology*, 27(3), 189-204.

[4] Y. Wu & W. Y. Zou, (1995) "Orthogonal frequency division multiplexing: a multi-carrier modulation scheme," *Consumer Electronics, IEEE Transactions on*, 41(3), 392-399.

[5] T. Jiang & Y. Wu, (2008) "An overview: peak-to-average power ratio reduction techniques for OFDM signals," *IEEE Transactions on broadcasting*, 54(2), 257.

[6] SE Alavi, IS Amiri, H Ahmad, ASM Supa'at & N Fisal, (2014) "Generation and Transmission of 3×3 W-Band MIMO-OFDM-RoF Signals Using Micro-Ring Resonators," *Applied Optics*, 53(34), 8049-8054.

[7] S. E. Alavi, I. S. Amiri, S. M. Idrus & A. S. M. Supa'at, (2014) "Generation and Wired/Wireless Transmission of IEEE802.16m Signal Using Solitons Generated By Microring Resonator," *Optical and Quantum Electronics*,

[8] S. Amiri, A. Nikoukar & J. Ali, (2013) "GHz Frequency Band Soliton Generation Using Integrated Ring Resonator for WiMAX Optical Communication," *Optical and Quantum Electronics*, 46(9), 1165-1177.

[9] C. K. Ho, Z. Lei, S. Sun & W. Yan, (2005) "Iterative detection for pretransformed OFDM by subcarrier reconstruction," *Signal Processing, IEEE Transactions on*, 53(8), 2842-2854.

[10] Abdolkarim Afroozeh, Iraj Sadegh Amiri, Alireza Zeinalinezhad & Harith Ahmad, *Characterization and Controlling of Soliton Signals Generated by Semiconductor Microring Resonators*. USA: Springer, 2015.

[11] S. Amiri, S. E. Alavi, Sevia M. Idrus, A. Nikoukar & J. Ali, (2013) "IEEE 802.15.3c WPAN Standard Using Millimeter Optical Soliton Pulse Generated By a Panda Ring Resonator," *IEEE Photonics Journal*, 5(5), 7901912.

[12] Iraj Sadegh Amiri & Abdolkarim Afroozeh, *Ring Resonator Systems to Perform the Optical Communication Enhancement Using Soliton.* USA: Springer, 2014.

[13] Y. Zhang, M. O'Sullivan & R. Hui, (2010) "Theoretical and experimental investigation of compatible SSB modulation for single channel long-distance optical OFDM transmission," *Optics express*, 18(16), 16751-16764.

[14] Y. Benlachtar, P. M. Watts, R. Bouziane, P. Milder, D. Rangaraj, A. Cartolano, R. Koutsoyannis, J. C. Hoe, M. Püschel & M. Glick, (2009) "Generation of optical OFDM signals using 21.4 GS/s real time digital signal processing," *Optics Express*, 17(20), 17658-17668.

[15] S. E. Alavi, I. S. Amiri, M. Khalily, A. S. M. Supa' at, N. Fisal, H. Ahmad & S. M. Idrus, (2014) "W-Band OFDM for Radio-over-Fibre Direct-Detection Link Enabled By Frequency Nonupling Optical Up-Conversion," *IEEE Photonics Journal* 6(6),

[16] B. J. Dixon, R. D. Pollard & S. Iezekiel, (2001) "Orthogonal frequency-division multiplexing in wireless communication systems with multimode fiber feeds," *Microwave Theory and Techniques, IEEE Transactions on*, 49(8), 1404-1409.

[17] Kim, Y. H. Joo & Y. Kim, (2004) "60 GHz wireless communication systems with radio-over-fiber links for indoor wireless LANs," *Consumer Electronics, IEEE Transactions on*, 50(2), 517-520.

[18] Y. Ma, Q. Yang, Y. Tang, S. Chen & W. Shieh, (2009) "1-Tb/s single-channel coherent optical OFDM transmission over 600-km SSMF fiber with subwavelength bandwidth access," *Optics express*, 17(11), 9421-9427.

[19] Iraj Sadegh Amiri & Harith Ahmad, *Optical Soliton Communication Using Ultra-Short Pulses.* USA: Springer, 2014.

[20] I. S. Amiri, M. R. K. Soltanian & H. Ahmad, Application of Microring Resonators (MRRs) in Soliton Communications, in

Optical Communication Systems: Fundamentals, Techniques and Applications, ed New York: Novascience Publisher, 2015.

[21] Sadegh Amiri, M. Nikmaram, A. Shahidinejad & J. Ali, (2013) "Generation of potential wells used for quantum codes transmission via a TDMA network communication system," *Security and Communication Networks*, 6(11), 1301-1309.

[22] D. Hillerkuss, R. Schmogrow, T. Schellinger, M. Jordan, M. Winter, G. Huber, T. Vallaitis, R. Bonk, P. Kleinow & F. Frey, (2011) "26 Tbit s-1 line-rate super-channel transmission utilizing all-optical fast Fourier transform processing," *Nature Photonics*, 5(6), 364-371.

[23] K. Takano, T. Murakami, Y. Sawaguchi & K. Nakagawa, (2011) "Influence of self-phase modulation effect on waveform degradation and spectral broadening in optical BPSK-SSB fiber transmission," *Optics express*, 19(10), 9699-9707.

[24] S. Amiri, S. E. Alavi & J. Ali, (2013) "High Capacity Soliton Transmission for Indoor and Outdoor Communications Using Integrated Ring Resonators," *International Journal of Communication Systems*, 28(1), 147–160.

[25] S. Amiri & H. Ahmad, (2015) "Multiplex and De-multiplex of Generated Multi Optical Soliton By MRRs Using Fiber Optics Transmission Link," *Quantum Matter*, 4(4),

[26] M. Nazarathy, J. Khurgin, R. Weidenfeld, Y. Meiman, P. Cho, R. Noe, I. Shpantzer & V. Karagodsky, (2008) "Phased-array cancellation of nonlinear FWM in coherent OFDM dispersive multi-span links," *Optics express*, 16(20), 15777-15810.

[27] Q. Yang, S. Chen, Y. Ma & W. Shieh, (2009) "Real-time reception of multi-gigabit coherent optical OFDM signals," *Optics express*, 17(10), 7985-7992.

[28] B. Djordjevic & B. Vasic, (2006) "Orthogonal frequency division multiplexing for high-speed optical transmission," *Optics Express*, 14(9), 3767-3775.

[29] K. Lee, C. T. Thai & J. K. K. Rhee, (2008) "All optical discrete Fourier transform processor for 100 Gbps OFDM transmission," *Optics express*, 16(6), 4023-4028.

[30] J. Lowery, (2010) "Design of arrayed-waveguide grating routers for use as optical OFDM demultiplexers," *Optics express*, 18(13), 14129-14143.

[31] W.-R. Peng, X. Wu, K. M. Feng, V. R. Arbab, B. Shamee, J. Y. Yang, L. C. Christen, A. E. Willner & S. Chi, (2009) "Spectrally efficient direct-detected OFDM transmission employing an iterative estimation and cancellation technique," *Optics express*, 17(11), 9099-9111.

[32] Iraj Sadegh Amiri, Ali Nikoukar & Sayed Ehsan Alavi, *Soliton and Radio over Fiber (RoF) Applications*. Saarbrücken, Germany: LAP LAMBERT Academic Publishing, 2014.

[33] S. Amiri & J. Ali, (2014) "Characterization of Optical Bistability In a Fiber Optic Ring Resonator," *Quantum Matter*, 3(1), 47-51.

[34] D. Hillerkuss, R. Schmogrow, M. Hübner, M. Winter, B. Nebendahl, J. Becker, W. Freude & J. Leuthold, (2010), "Software-defined multi-format transmitter with real-time signal processing for up to 160 Gbit/s", in *Signal Processing in Photonic Communications* SPTuC4.

[35] Y. S. Neo, S. M. Idrus, M. F. Rahmat, S. E. Alavi & I. S. Amiri', (2014) "Adaptive Control for Laser Transmitter Feedforward Linearization System," *IEEE Photonics Journal* 6(4),

[36] S. E. Alavi, I. S. Amiri, S. M. Idrus, ASM Supa'at, J. Ali & P. P. Yupapin, (2014) "All Optical OFDM Generation for IEEE802.11a Based on Soliton Carriers Using MicroRing Resonators," *IEEE Photonics Journal*, 6(1),

[37] IS Amiri, SE Alavi, N Fisal, ASM Supa'at & H Ahmad, (2014) "All-Optical Generation of Two IEEE802.11n Signals for 2×2 MIMO-RoF via MRR System," *IEEE Photonics Journal*, 6(6),

[38] S. Amiri, S. E. Alavi, S. M. Idrus, A. S. M. Supa'at, J. Ali & P. P. Yupapin, (2014) "W-Band OFDM Transmission for Radio-over-

Fiber link Using Solitonic Millimeter Wave Generated by MRR", *IEEE Journal of Quantum Electronics*, 50(8), 622 - 628.

[39] D. Hillerkuss, M. Winter, M. Teschke, A. Marculescu, J. Li, G. Sigurdsson, K. Worms, S. Ben Ezra, N. Narkiss & W. Freude, (2010) "Simple all-optical FFT scheme enabling Tbit/s real-time signal processing," *Optics express*, 18(9), 9324-9340.

[40] K. Sathananthan & C. Tellambura, (2001) "Probability of error calculation of OFDM systems with frequency offset," *Communications, IEEE Transactions on*, 49(11), 1884-1888.

[41] C. C. Wei & J. J. Chen, (2010) "Study on dispersion-induced phase noise in an optical OFDM radio-over-fiber system at 60-GHz band," *Optics express*, 18(20), 20774-20785.

[42] Iraj Sadegh Amiri, Abdolkarim Afroozeh & Harith Ahmad, *Integrated micro-ring photonics: Principles and Applications as Slow light devices, Soliton generation and optical transmission.* United States: CRC Press, 2015.

[43] Iraj Sadegh Amiri, Sayed Ehsan Alavi, S. M. Idrus, Abdolkarim Afroozeh & Jalil Ali, *Soliton Generation by Ring Resonator for Optical Communication Application.* New York: Nova Science Publishers, 2014.

[44] Iraj Sadegh Amiri, Sayed Ehsan Alavi & Sevia Mahdaliza Idrus, *Soliton Coding for Secured Optical Communication Link.* USA: Springer, 2014.

[45] S. Amiri, R. Ahsan, A. Shahidinejad, J. Ali & P. P. Yupapin, (2012) "Characterisation of bifurcation and chaos in silicon microring resonator," *IET Communications*, 6(16), 2671-2675.

[46] S. Amiri, M. Ebrahimi, A. H. Yazdavar, S. Gorbani, S. E. Alavi, Sevia M. Idrus & J. Ali, (2014) "Transmission of data with orthogonal frequency division multiplexing technique for communication networks using GHz frequency band soliton carrier," *IET Communications*, 8(8), 1364 – 1373.

[47] Afroozeh, I. S. Amiri, M. A. Jalil, M. Kouhnavard, J. Ali & P. P. Yupapin, (2011) "Multi Soliton Generation for Enhance Optical Communication," *Applied Mechanics and Materials*, 83 136-140.

[48] S. Amiri, K. Raman, A. Afroozeh, M. A. Jalil, I. N. Nawi, J. Ali & P. P. Yupapin, (2011) "Generation of DSA for security application," *Procedia Engineering*, 8 360-365.

[49] S. Amiri & J. Ali, (2014) "Optical Quantum Generation and Transmission of 57-61 GHz Frequency Band Using an Optical Fiber Optics," *Journal of Computational and Theoretical Nanoscience (CTN)*, 11(10), 2130-2135.

Chapter 2

PRINCIPLE OF ALL-OPTICAL OFDM TRANSMISSION SYSTEMS

ABSTRACT

In this chapter, theory and principle of All-Optical OFDM transmission systems are presented. Implementing orthogonal frequency division multiplexing (OFDM) makes it doable to succeed in terabit per second (T bps) single-channel line rates so such a speed now not appears to be too far-fetched. In wavelength division multiplexing (WDM), several wavelengths carry several data streams in parallel. Spectral guard bands are required to avoid crosstalk from one channel to another. OFDM could be a special category of multi-carrier modulation (MCM). A fundamental challenge with OFDM is that a lot of sub-channels are needed so that the transmission channel has effects on each sub-carrier to be a flat channel. Pretty much linearly with many sub-channels N. Second, many orthogonal sub-channels are generated and demodulated with no resorting too much more complex RF oscillator and filters. Howell identifies fast Fourier transform (FFT) that way the FFT can be an efficient method to help calculate the distinct Fourier transform (DFT) for several times samples N. Nonlinearities were created when the power of electromagnetic wave incident substantially increases. The optical fiber medium is only able to be believed just like a linear medium when the launch power is sufficiently low. Infused silica, the nonlinear effects in optical materials might be observed at relatively low levels of power.

For our discussion of phase noise of incident light, first, we need to consider characteristic of the medium to drive the relative wave equation.

Keywords: multicarrier modulation, wave equation, discrete fourier transform, intersymbol interference, coupler-based optical IFFT/FFT

2.1. INTRODUCTION TO OFDM

OFDM could be a promising technology in wireless and optical transmission system's space to hold high-data rates. Implementing OFDM makes it doable to succeed in terabit per second (T bps) single-channel line rates so such a speed now not appears to be too far-fetched. Historically, time-division multiplexing (TDM), within which multiple bitstreams multiplexed onto one signal by assigning a recurrent time slot to each of the tributaries, has accumulated the amount of data encoded on single laser wavelength. The diagram of TDM is shown in Figure 2.1. Recently, data streams of 10.2 T bps are encoded on a single laser by applying a TDM transmission system. However, TDM pulses are short and spectrally broad as illustrated in Figure 2.3(a) [1-5].

This makes TDM schemes difficult to implement. The short pulses need a narrow receiver time window, and therefore the large optical bandwidth (for example 30nm) makes exactly designed dispersion compensation a necessity.

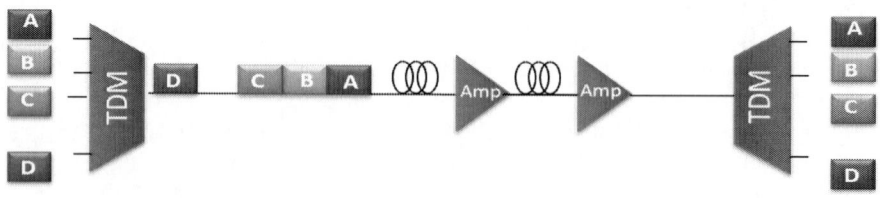

Figure 2.1. Simplified optical TDMA transmission system.

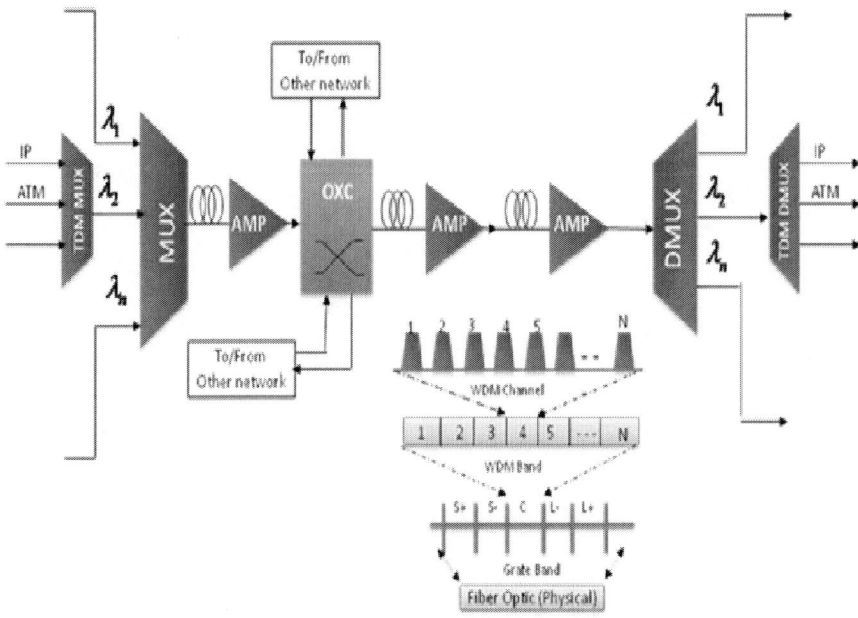

Figure 2.2. WDM transmission system.

Alternatively, with WDM, high data bitrates transmitted over optical fibers. In WDM, a number of wavelengths carry several data streams in parallel. A typical WDM optical transmission system is illustrated in Figure 2.2. The duration of pulses is longer and optical bandwidth is moderated as shown in Figure 2.3(b).

Spectral guard bands are required to avoid crosstalk from one channel to another. This reduces spectral efficiency, and the dedicated receivers and transmitters with stabilized lasers for each transmission wavelength make the systems expensive. In contrast to WDM, modulated OFDM sub-channels overlap each other significantly due to orthogonally of subcarriers [6-9]. Spectral efficiency is therefore high, and the encoding of information on a large number of sub-channels makes OFDM tolerant towards dispersion. Figure 2.3(c) illustrates the features of OFDM signal.

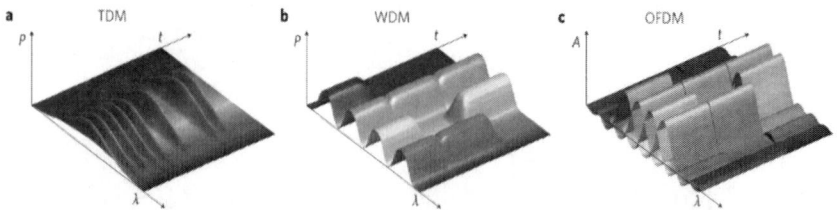

Figure 2.3. Comparison of multiplexing schemes in optical transmission systems; a) Data are transmitted in the form of a serial stream of pulses. b) For WDM, data are distributed over wavelengths and transmitted in parallel. The different wavelength tributaries can be separated by optical bandpass filters. c) In an OFDM modulation scheme, data are transmitted on a number of subcarriers in parallel.

2.1.1. The Basic Principle of OFDM

OFDM could be a special category of Multi-Carrier Modulation (MCM), generic implementation of that is delineated in Figure 2.4.

The structure of a complex multiplier (IQ modulator/demodulator), that is often utilized in MCM systems, is additionally shown in the figure. The MCM transmitted signal s(t) is described as :

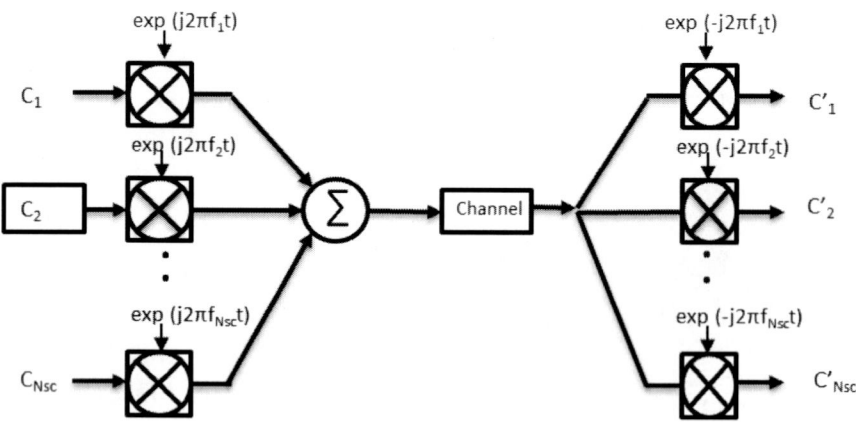

Figure 2.4. Diagram for a generic multicarrier modulation.

$$s(t) = \sum_{i=-\infty}^{+\infty} \sum_{k=1}^{N_{sc}} c_{ki} s_k(t - iT_s) \tag{2.1}$$

$$s_k(t) = \prod(t) e^{j2\pi f_k t} \tag{2.2}$$

$$\prod(t) = \begin{cases} 1, 0 < t \leq T_s \\ 0, t \leq 0, t \geq T_s \end{cases} \tag{2.3}$$

is the i-th data image at the k-th subcarrier, s_k is that the wave equation for the k-th sub-channel, N_{sc} is that the range of sub-channels, f_k is that the frequency of the subcarrier, T_s is that the period, and $\prod(t)$ is that the pulse shaping equation. The optimum detector for every sub-channel may use a filter that matches the sub-channel waveform or a correlator matched to carrier the the as shown in Figure 2.4. Therefore, the detected data image c_{ik}' at the correlator is given by

$$c_{ik}' = \frac{1}{T_s} \int_0^{T_s} r(t - iT_s) \, s_l^* dt = \frac{1}{T_s} \int_0^{T_s} r(t - iT_s) e^{-j2\pi f_k t} \, dt \tag{2.4}$$

where r (t) is that the received time domain signal. The classical MCM uses non-overlapped band-limited signals and may be applied with a bank of vast numbers of oscillators and filters at each transmit and receive end, the foremost disadvantage of MCM is large bandwidth. Bandwidth because of designing filters and oscillators cost-effectively, the channel spacing needs to be a multiple of the symbol rate, greatly reducing the spectral efficiency. A unique approach, OFDM, was investigated by using overlapped orthogonal signal sets. This orthogonality originates from a correlation between any two subcarriers, given by

$$\delta_{kl} = \frac{1}{T_s} \int_0^{T_s} s_l^* s_k = \frac{1}{T_s} \int_0^T \exp(j2\pi(f_k - f_1)t) \, dt = \exp(j\pi(f_k - f_1)T_s) \frac{\sin(\pi(f_k - f_1)T)}{\pi(f_k - f_1)T} \tag{2.5}$$

Obviously, it can be said that if condition

$$f_k - f_1 = m\frac{1}{T_s} \tag{2.6}$$

will be satisfied, then the couple subcarriers are orthogonal together. This signifies the orthogonal sub-channel sets, with their frequencies spaced at multiples of the inverse of the symbol periods, might be recovered with the matched filters with Eq. (2.4) with no inter-carrier interference (ICI), even with strong signal overlapping.

2.1.2. Discrete Fourier Transform Implementation of OFDM

A fundamental challenge with OFDM is that a lot of sub-channels are needed so that the transmission channel has effects on each sub-carrier to be a flat channel. This contributes to an extremely difficult architecture involving many oscillators and filters at both transmit and receive concludes. To cope whit this problem, inverse distinct Fourier transforms (IDFT)/distinct Fourier transform (DFT) in OFDM modulation/demodulation was employed. Eq. (2.1) and (2.4) represent OFDM modulation and OFDM demodulation, respectively. Temporarily disregard the index i, re-denote N_{sc} since N in Eq. (2.1) to focus our attention using one OFDM symbol, and assume that we sample s(t) with every interval connected with T_s/N. The mth sample of s(t) from (Eq. 2.1) becomes.

$$s_m = \sum_{k=1}^N c_k \cdot e^{j2\pi f_k \frac{(m-1)T_s}{N}} \tag{2.7}$$

Using the orthogonality condition of Eq. (2.6) and the conversion that

$$f_k = \frac{k-1}{T_s} \tag{2.8}$$

And replacing (Eq.2.7) into (Eq.2.8), we have

$$s_m = \sum_{k=1}^{N} c_k \cdot e^{j2\pi f_k \frac{(m-1)T_s}{N}} = \sum_{k=1}^{N} c_k \cdot e^{j2\pi \frac{(m-1)(k-1)T_s}{N}} = \Im^{-1}\{c_k\}$$

(2.9)

where $\Im$ is the Fourier transform, and $m \in [1, N]$. In a similar manner, at the receive end, we arrive at

$$c_k' = \Im\{r_m\}$$

(2.10)

where r_m would be the received sampled from every interval of T_s/N. From Eqs. (2.9) in addition to (2.10), it can be shown that the transmitted OFDM signal s(t) is merely a simple N-point IDFT in the information symbol c_k, and also the received information symbol c_k' is a simple N-point DFT of the receive sampled signal. It is worth noting we now have two critical devices have assumed for DFT/IDFT using in OFDM. First, because of the existence of an efficient of a powerful IFFT/FFT algorithm, a lot of complex multiplications intended for IFFT in Eq. (2.9) and FFT in Eq. (2.10) is reduced from N^2 to

$$\frac{N}{2}\log_2 N$$

(2.11)

Pretty much linearly with many sub-channels N. Second, many orthogonal sub-channels is generated and demodulated with no resorting too much more complex RF oscillator and filters. This leads to a relatively simple structure for OFDM implementation when a lot of sub-channels are needed. The corresponding structure using DFT/IDFT along with DAC/ADC is demonstrated in Figure 2.5. At this transmitter, the input serial data bits are first changed into many parallel data pipes, each mapped onto corresponding information symbols for the sub-channels within one particular OFDM symbol, and digital time domain signal is obtained through the use of IDFT, which is subsequently inserted having a guard interval and changed into real-time waveform by means of DAC. The guard period of time is inserted to prevent intersymbol interference (ISI) due to channel dispersion. At the baseband receiver, the OFDM signal is down transformed into baseband with the IQ demodulator, sampled having an ADC, and then demodulated by performing DFT along with baseband signal processing to recover the data.

Through Eq. (2.7), it really is noted that the actual OFDM signal s_m is a periodical function of f_k using a period of N/T_s. Thus, in Eq. (2.7) in addition to Eq. (2.8), the sub-channel frequency f_k as well as index k is usually generalized as.

$$f_k = \frac{k-1}{T_s} , k \in [k_{min} + 1, k_{min} + N] \tag{2.11}$$

where k_{min} is a random integer. However, only two sub-channel index conventions are widely used: $k \in [1, N] \ and \ k \in \left[-\frac{N}{2} + 1, \frac{N}{2} \right]$. These two conventions are mathematically equivalent.

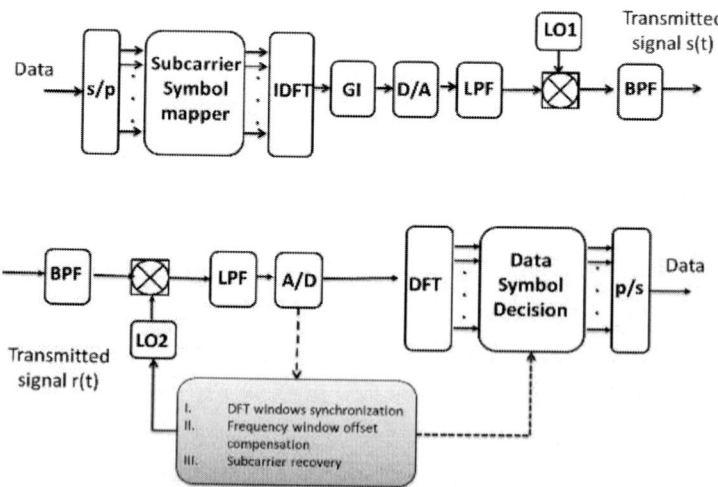

Figure 2.5. Conceptual diagram for (a) OFDM transmitter (b) OFDM receiver.

2.2. ALL-OPTICAL OFDM

Within conventional OFDM, both FFT/IFFT are typically performed in the electronic domain and for that reason limited in touch rate. Now, real-time electronic IFFT in addition to FFT signal digesting for OFDM signals nearly 101.5 Gbps has become demonstrated. This limitation seems to be too far-fetched to succeed in desirable for the generation or reception of terabit per second OFDM signal. An all-optical solution which may work beyond the state-of-art technology speed would, therefore, be of interest [10-13].

2.2.1. IFFT/FFT

Howell identifies Fast Fourier Transform (FFT) that way The FFT can be an efficient method to help calculate the distinct Fourier transform (DFT) for a number of times samples N, where $N = 2^p$ with p just as one integer. The N point DFT is given as

$$X_m = \sum_{n=0}^{N-1} e^{-j2\pi mn/N} x_n , \quad m=0, 1, \ldots, N-1 \qquad (2.12)$$

for transforming the N inputs of x_n into N outputs associated with X_m. If the X_n indicates a time-series associated with equidistant signal samples of signal x(t) over the time period T then the X_m will be the unique complex spectral components of signal x repeated with period T. the FFT normally decimates a DFT associated with size N into two interleaved DFTs associated with size N/2 in several recursive stages to ensure that.

$$X_m = \begin{cases} E_m + \exp\left[-j\,\frac{2\pi}{N}m\right] O_m, & m < N/2 \\ E_{m-N/2} - \exp[-j\,\frac{2\pi}{N}(m - \frac{N}{2})]O_{m-\frac{N}{2}}, & m \geq N/2 \end{cases} \qquad (2.13)$$

The quantities E_m and O_m are the even and odd DFT of size N/2 for even and odd inputs x_{2l} and x_{2l+1} (l = 0,1,2,...,N/2-1), respectively.

2.2.2. Coupler - Based Optical IFFT/FFT

Mahric has demonstrated a possible implementation of the optical circuit which performs as FFT [14]. Take into account the N×N single-mode star coupler of Figure 2.6. The fields injected at the inputs a_i are coherent, since they are derived from an identical monochromatic source, and the output fields b_j results from the interference among the input signals as they are split and superimposed by the individual elements making up the N×N star. An ideal lossless star coupler is such that

$$b_j = \frac{1}{\sqrt{N}} \sum_{i=0}^{N} \epsilon_{ji} a_i \qquad (2.14)$$

where,

$$|\epsilon_{ji}| = 1 \qquad (2.15)$$

Eqs. (2.14) as well as (2.15) indicate that, if power is incident at only one input, it emerges evenly divided among all N inputs. It is valid for any choice of the unit-magnitude complex phase factors ϵ_{ji}.

If power is now incident at two or more inputs, the output powers are, generally, no longer even because of interference resulting from the values of ϵ_{ji}. The types of discrete transform which are obtained, i.e., the kinds of discrete transform which are represented by Eq. (2.14), are thus determined by the set of ϵ_{ji}. There is some degree of arbitrariness in the choice of these phase factors, although they need to satisfy the power conservation equation

$$\sum_{j=1}^{N} b_j b_j^* = \sum_{i=1}^{N} a_i a_i^* \qquad (2.16)$$

By bearing in mind that the Eq. (2.15), the Eq. (2.16) it can be modified as

$$\frac{1}{N} \sum_{k=1}^{N} \sum_{l=1}^{N} a_k a_l^* \sum_{j=1}^{N} \epsilon_{jk} \epsilon_{jl}^* = \sum_{i=0}^{N} a_i a_i^* \qquad (2.17)$$

The Eq. (2.17) would be desirable if

$$\sum_{j=1}^{N} \epsilon_{j,k} \epsilon_{j,1}^* = N\delta_{k,l} \qquad (2.18)$$

where $\delta_{k,l}$ is the Kronecker delta. There are a lot of possible sets of ϵ_{ji} that fulfill Eq. (2.18). A particularly interesting one is obtained from the Nth roots of unity, namely, for

$$\epsilon_{jk} = \exp(i2\pi jk/N)\, a_k \qquad (2.19)$$

In that case, we now have that

$$b_j = \frac{1}{\sqrt{N}} \sum_{k=1}^{N} \exp\left(\frac{i2\pi jk}{N}\right) a_k \qquad (2.20)$$

Definitely the Eq. (2.20) states the actual output spatial pattern would be the Discrete Fourier Transform (DFT) of the input spatial pattern. Therefore, in optics, DFT operation theoretically evaluated utilizing passive optical networks. For instance, FFT/IFFT of the discrete array amplitudes could be implemented by interferometry utilizing asymmetric couplers.

In practice, a 2×2 asymmetric coupler is needed that its output amplitude is the same as a 2×2 FFT:

$$S_0 = \frac{1}{\sqrt{2}} (s_0 + s_1),\ S_1 = \frac{1}{\sqrt{2}} (s_0 - s_1) \qquad (2.21)$$

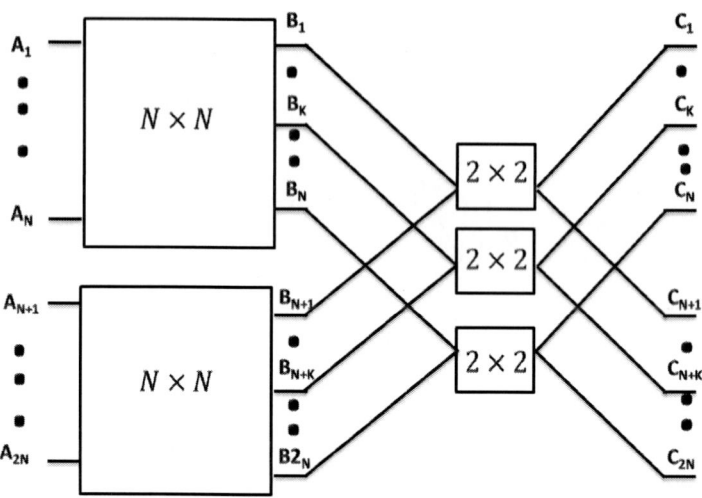

Figure 2.6. A 2N × 2N star coupler.

It's possible to derive and implement a Nth-order FFT through making proper interconnection between two (N/2)th-order FFT blocks, N/2 second-order FFTs, and appropriate phase shift elements. The first N/2 outputs are :$(0 \leq m \leq \frac{N}{2} - 1)$

$$S_m = \frac{1}{\sqrt{2}} \left\{ \frac{1}{\sqrt{N/2}} \sum_{n=0}^{\frac{N}{2}-1} S_{2n} \cdot e^{\frac{-jmn2\pi}{N/2}} + e^{\frac{-jm2\pi}{N}} \cdot \frac{1}{\sqrt{N/2}} \sum_{n=0}^{\frac{N}{2}-1} S_{2n+1} \cdot e^{\frac{-jmn2\pi}{N/2}} \right\}$$

(2.22)

The first expression shows the matching inputs from our first (N/2) th-order FFT block while the second expression shows the methods from our second block having added phase shift -$\frac{m2\pi}{N}$. As it is already talked about the Eq. (2.22) refers to the first N/2 outputs but the remaining N/2 outputs could be derived using similar relation. Intended for IFFTs, just one simply should change the signs of the phase shift inversely in comparison with the FFT. This mapping on the passive coupler network determined by Eq. (2.22) will generate the familiar butterfly structure, while the interconnections along with the phase shifters within a fraction of the wavelength to maintain proper interferometer operation. Truly, FFT is a successful way to calculate the discrete Fourier transform for a number of period samples N, where $N = 2^p$ having p an integer. This N-point DFT will be granted since

$$X_m = \sum_{n=0}^{N-1} \exp\left[-j2\pi \frac{mn}{N}\right] x_n \,, m = 0, \dots, N-1 \tag{2.23}$$

transforming the N inputs x_n into N outputs X_m. whenever the x_n represents a time-series of equidistant signal x(t) on the time period T, as shown in Figure 2.7(a), then the X_m will be the unique complex spectral components of signal x repeated with period T. Figure 2.7(b) demonstrates the direct implementation on the the FFT pertaining to N = 4 applying time electrical sampling and signal processing. Mahric and Siegman suggested optical FFT As shown in Figure 2.9(c) [15-17].

Figure 2.7(c) demonstrates the direct implementation of the optical FFT is regarding N = 4 using time electrical sampling. When using an optical circuit to calculate the FFT the outputs X_m appear instantaneously for any given input combination x_n. Since so as to extract the spectral components of a time series the N time samples in interval T need to be fed simultaneously to the circuit. This can be achieved using optical time delays being a serial-to-parallel (S/P) convertor, illustrated in Figure 2.7(c). Optical FFT contrary to its electronics counterpart has the continuous mode of operation. To be more obvious, in the electronic domain, the optical signal is sampled along with the FFT is computed from all examples x_n. Then, the following N samples are taken. In the optical domain, The FFT computed continuously.

In fact, only when feeder outlines 1 to N contain the time samples from within their specific interval, in this case, the calculation is correct. Sampling must, as a result, be performed in synchronization with the symbol over the duration of T/N. Subsequently, these samples can be processed in the optical FFT stage. However, it needs to be taken into account that proper calculation is only possible if all samples are forwarded from one stage to the next stage in synchronism. To meet FFT conditions, in addition to aforementioned requirements, couplers need to have equal delay waveguides and also keep proper phase relations as shown in Figure 2.7(c).

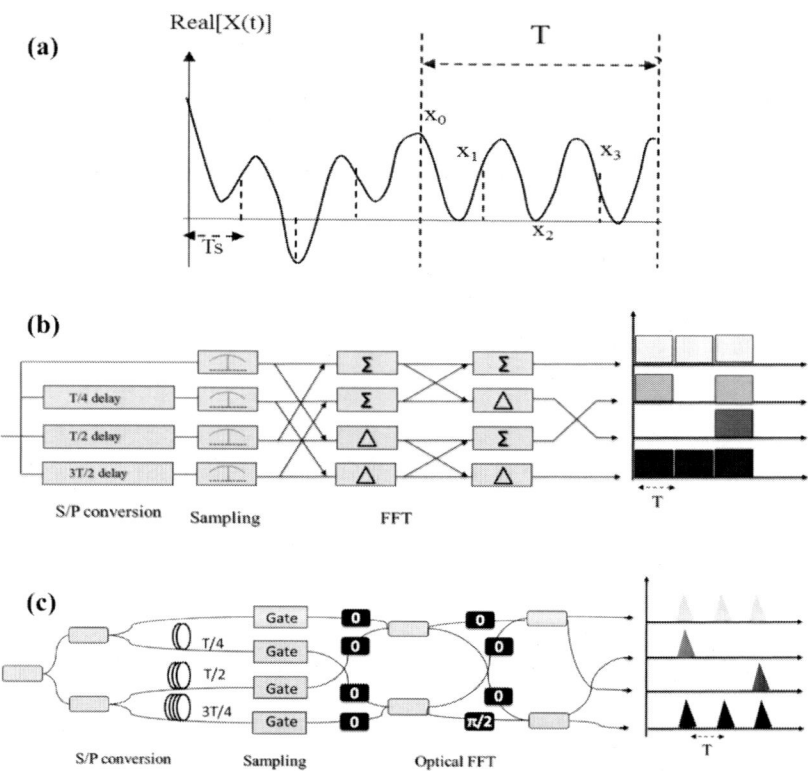

Figure 2.7. Example of the traditional four-point fast Fourier transform and its optical equivalent. (a) The exemplary signal in time sampled at $N = 4$ points. (b) The structure contains a serial-to-parallel (S/P) conversion which makes parallel samples of the signal, a selection stage to generate the time sample x_n and a conventional FFT stage that calculates the fast Fourier form of the sampled signal. (c) The optical equivalent of the circuit uses passive splitters and optical time delays for serial-to-parallel conversion; optical gates perform the sampling of the optical waveform; afterward, the optical FFT is computed using optical 2×2 couplers and phase shifts.

However, the proposed FFT suffers from complexity with increasing the quantity (N) of FFT, because the number of couplers stems from the complexity $C_{std} = N-1 + (N/2)\log_2(N)$, and the optical phases in all $N \log_2(N)$ arms of the FFT structure must be stabilized regarding each other, thereby limiting N to small number for practical cases. This hinders proposed optical FFT structure to be practical. Electronic signal processing is alternative to optical one. Meanwhile, it is limited due to its power consumption and its restricted speed. Hillerkuss streamline the circuit of Figure 2.7(c) without affecting its operation [18]. Figure 2.8 reveals that re-ordering the delays and through re-marking the outputs consequently an equivalent but simpler implementation are available, hence a primary implementation of the circuit in Figure 2.7(c) could be difficult to make for its frequent waveguide crossings, and due to the many waveguide phases that should be accurately controlled.

The simplifying steps as an example with N = 4 are in Figure 2.8. At a starting point, the sampling gates are relocated towards the end of the circuit. It will consider the re-ordering the structure of the standard FFT will not change the function of it. Next, the delays in the S/P transformation stage was reordered as indicated in Figure 2.8(a) and re-mark the outputs accordingly. In this implementation, OFFT contains two parallel delay interferometers (DIs) with the exact same free spectral range (FSR) but different absolute delays are shown in 2.8(b). By moving the common delay of T/4 in both arms of the lower DI to its outputs, one gets to the same DIs using the same input signal, see Figure 2.8(c). This redundancy can be eliminated by replacing both the DIs with one DI and by splitting the result which is displayed in Figure 2.8(d). These simplifications can be distributed to FFTs for any size N. This new optical FFT processor consists only of N-1 cascaded DIs with a small complex of only $C_{DI} = 2(N-1)$ couplers by which $C_{DI} \leq C_{sid} \forall N$. Also, with this implementation simply the phase of N-1 DIs requirements stabilization and no inter-DI phase adjustment is essential.

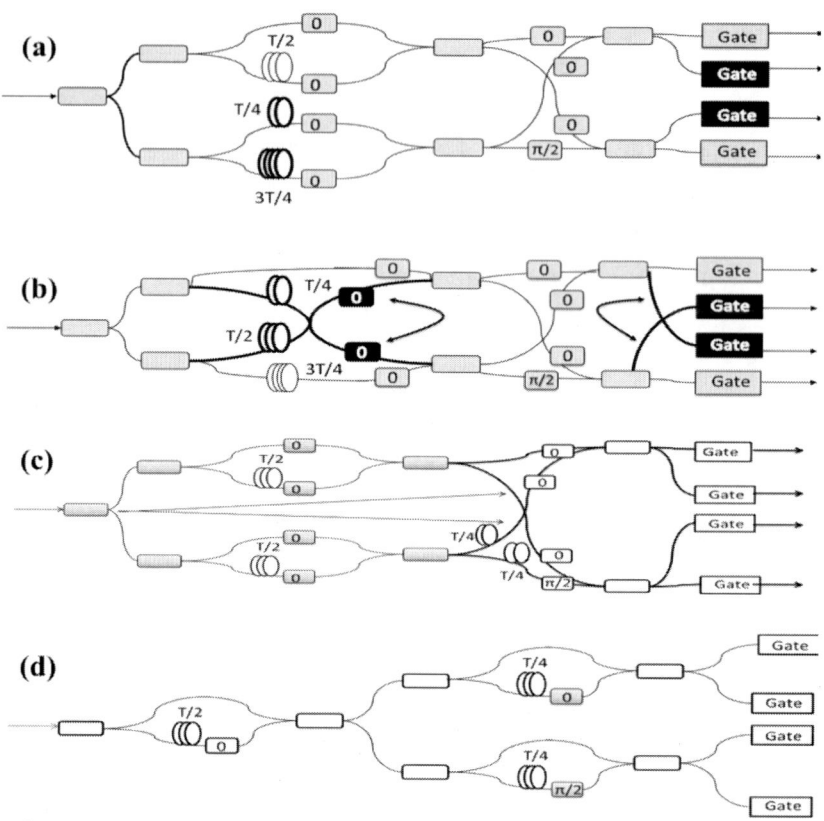

Figure 2.8. Four-point optical FFT for symbol period T; (a) traditional implantation; (b) leading to a structure consisting of two Dis with the same differential delay; the additional T/4 delay is moved out of the second DI (c), which leads to two identical DIs that can be replaced by a single DI followed by signal splitters; (d) low-complexity scheme with combined S/P conversion and FFT.

As outlined in Figure 2.8, it is obvious that this DI delay in each stage and the location of the required phase shifters is extracted at the same time. Figure 2.9(a) shows that conventional recommended structure based on Mahric and Figure 2.9(b) demonstrates the new structure according to for 8-stage FFT [18].

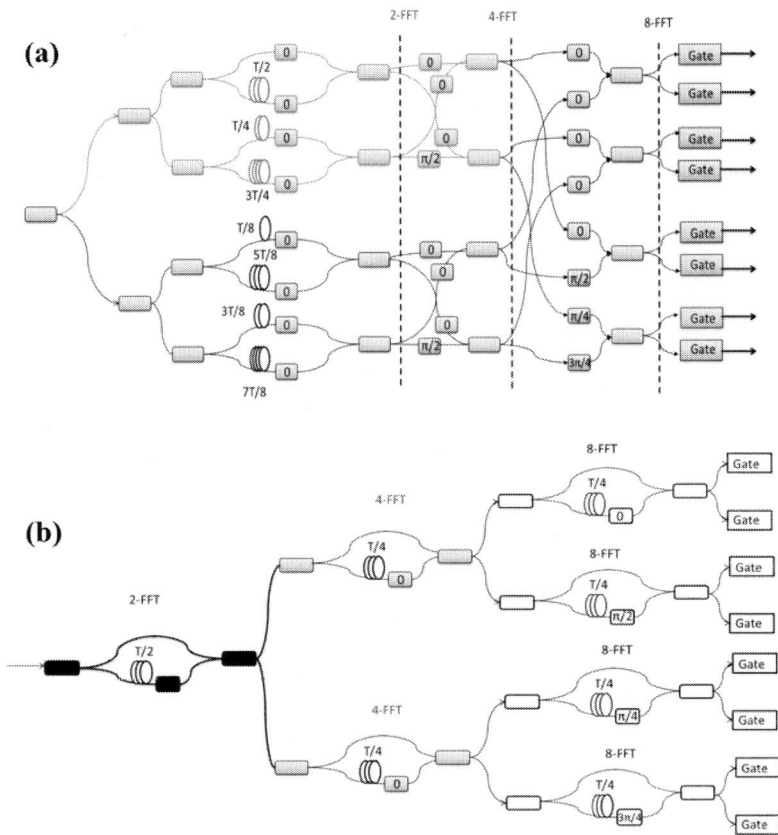

Figure 2.9. (a) Direct FFT application versus (b) refined all-visual FFT circuit for N = 8 showing the layout of delays and phase shifts.

An important advantage of this approach is the fact for a single frequency component of the sampled signal, other parts of the new proposed structure are not included. Actually, by tuning the phases in each DI, any arbitrary FFT coefficient of the signal can be selected without having changed the structure in the setup. While it is impossible to achieve this type of convenient with the typical method shown in Figure 2.8(a) and Figure 2.9(a). It is also important to know how DFT/FFT is equivalent to a cascade of DI. Actually, FFT has the frequency response from the DIs. The DFT can be written for continuous output and input signals b x(t) and Xm(t) as

$$X_m(t) = \frac{1}{N} \cdot \sum_{n=0}^{N-1} \exp\left(-j2\pi n \frac{m}{N}\right) \cdot \delta\left(t - n\frac{T}{N}\right) * x(t) \qquad (2.24)$$

where $\delta(\ )$ is the Dirac delta function used to sample the input signal x(t) at N equidistant points within the interval T by means of the convolution operation (*). As a result of numerous convolutions taking place when calculating the impulse response of cascade element, it can be more appropriate to do comparison in the frequency domain. The transfer function $H_m(\omega)$ to the DFT can be obtained with a Fourier transform of (Upper)

$$\hat{X}_m(\omega) = \frac{1}{N} \sum_{n=0}^{N-1} \exp\left(-j2\pi \frac{nm}{N}\right) \exp\left(-j\omega \frac{nT}{N}\right) \cdot \hat{x}(\omega) \qquad (2.25)$$

In which the caret (^) indicates the Fourier transform. It could be seen in two sums corresponding to even and unusual n,

$$H_m(\omega) = \frac{1}{N} \sum_{n=0}^{\frac{N}{2}-1} \left\{ \exp\left(-j2\pi[2n]\frac{m}{N}\right) \exp\left(-j\omega[2n]\frac{T}{N}\right) + \right.$$
$$\left. \exp\left(-j2\pi[2n+1]\frac{m}{N}\right) \exp\left(-j\omega[2n+1]\frac{T}{N}\right) \right\} \qquad (2.26)$$

Which could be simplified to

$$H_m(\omega) = \frac{2}{N} \cdot \sum_{n=0}^{\frac{N}{2}-1} \exp\left(-j\frac{2n}{N}[2\pi m + \omega T]\right) \cdot \frac{1}{2}[1 + \exp(-j[\omega\frac{T}{N} + \frac{2\pi m}{N}])] \qquad (2.27)$$

As shown, $H_{pm,1}$ will be the n-independent transfer function for the upper input of a delay interferometer with delay

$$T_p = \frac{T}{N} = \frac{T}{2^p} \qquad (2.28)$$

And additional phase shift

$$\varphi_{pm} = 2\pi \frac{m}{N} - \pi \tag{2.29}$$

From the cascade of directional couplers and a delay line in the lower arm,

$$H_{pm}(\omega) = \frac{1}{\sqrt{2}} \begin{pmatrix} 1 & j \\ j & 1 \end{pmatrix} \cdot \begin{pmatrix} 1 & 0 \\ 0 & \exp[-j(\omega T_p + \varphi_m)] \end{pmatrix} \cdot \frac{1}{\sqrt{2}} \begin{pmatrix} 1 & j \\ j & 1 \end{pmatrix} \begin{pmatrix} 1 & 0 \\ 0 & 0 \end{pmatrix} \tag{2.30}$$

The expression (*) consists of two major parts including the transfer function of the DFT of order N/2 and also transfer function of the delay line and coupler. For this reason, a DFT of order N, with $N = 2^p$, can be implemented optically by cascading a DFT of order N/2 and a delay interferometer with delay T/N and output-specific phase shift φ_{pm}. It can be easily verified how the DFT transfer work for N = 2 is equal to both outputs of a one DI.

The DI phase φ_{pm} for an upper arm output X_m, is the same as that for the lower arm result $jX_{m+N/2}$. The term describing the N/2-order FFT in (*) is also the same for X_m and $jX_{m+N/2}$ due to its periodicity. Thus both outputs of the single DI could be used to obtain different coefficients of the DFT, resulting directly in the visual FFT scheme of Figure 2.9.

Sensitivity of all-optical OFDM system with phase φ_{pm} shows that the system suffer from Phase noise effects and also being more sensitive to Carrier Frequency Offset (CFO) which creates Phase Rotate Term (PRT) on each subcarrier and result in inter-carrier interference (ICI) due to destroying orthogonality of subcarriers In fiber-optic communication systems, abovementioned weaknesses of OFDM might result in higher sensitivity to fiber nonlinearities such as self-phase modulation (SPM), cross-phase modulation (XPM) and four-wave mixing (FWM). Therefore, precisely calculation of induced nonlinearity, which is produced by fiber dispersion, is crucial to cope with these side effects of OFDM.

2.3. FIBER OPTIC NONLINEARITY EFFECTS

In fiber optic communication techniques, linear impairments are due to the fiber loss, chromatic dispersion (CD) and polarization mode dispersion (PMD) [19-25]. Optical power loss due to light propagation inside the fiber results from the absorption and scattering and it can easily be compensated by optical amplifiers. CD and PMD would be the main linear impairments for optical interaction systems. Other impairments are due to fiber nonlinearity such as self-phase modulation (SPM) [26-27], cross-phase modulation (XPM), 4-wave mixing (FWM), stimulated Raman Scattering (SRS) and Stimulated Brillouin Scattering (SBS). The nonlinear impact in an optical fiber may be categorized into two principal categories as demonstrated in Figure 2.10.

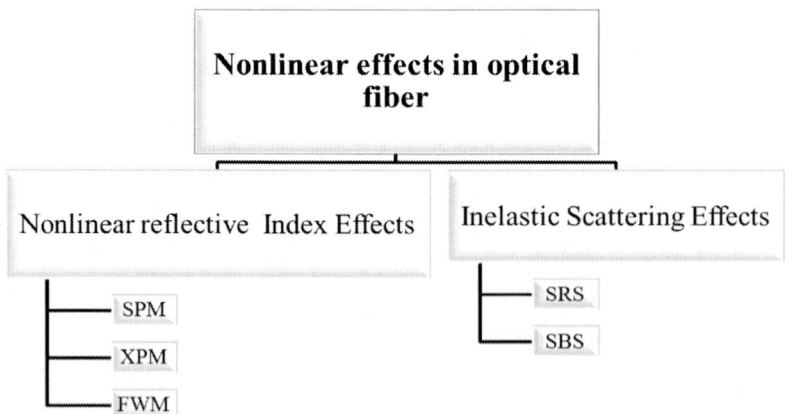

Figure 2.10. Classification of nonlinear effects in optical fiber.

Nonlinearities were created when the power of electromagnetic wave incident substantially increases. The optical fiber medium is only able to be believed just like a linear medium when the launch power is sufficiently low. For long-haul fiber optic transmission system and wideband WDM systems, to combat accumulated noise added to the amplifier chain over the transmission fiber link, the launch power must to be increased to keep the signal to noise ratio (SNR) sufficient for error-free detection at the receiver. Since the launch power increases, the nonlinearity of fiber has become significant and lead to low performance. Nonlinear effects in optical fibers are generally due to two causes. The initial event is dependent on the fact the index of refraction of several materials, including glass, related to the light intensity [28-34]. The main origins of this from nonlinear response come from the harmonic motion of bound electrons under the influence of an applied field. This phenomenon is called the Kerr effect that is discovered in 1875 by John Kerr [35-37]. The second event cause is a non-elastic scattering of photons in optical materials, which in results in stimulated Raman and stimulated Brillouin scattering phenomena. Furthermore, for the dependence in the index of refraction with wavelength, this leads to dispersion effects [37-42]. The refractive index might be written as

$$n(\omega, P) = n_0(\omega) + n_2 \frac{P}{A_{eff}} \tag{2.31}$$

where n_0 the linear part of the refractive index, n_2 is Kerr coefficient with typical value of 2.2 - 3.4 $\times$ 10^{-20} m^{-20}/W, P is the optical power, and A_{eff} is the effective core area [43-50]. In fiber optics, the Kerr coefficient is insignificant compared to the amount of other nonlinear media by no less than two orders of magnitude. Despite, the naturally small values in the nonlinear coefficients in fused silica, the nonlinear effects in optical materials might be observed at relatively low levels of power. It is possible because of two important characteristics of single mode fiber (SMF) first a little effective core area and second extremely low loss (0.2 dB/km). The dependence of the refractive index on the light intensity results in the propagation constant, β varying as the light intensity due to $= 2\pi n/\lambda$, and the propagation constant can be written as

$$\beta(\omega, P) = \beta_0(\omega) + \frac{2\pi n_2}{\lambda A_{eff}} \qquad (2.32)$$

where $\beta_0(\omega)$ is the propagation constant in the absence of nonlinear effects, and

$$\gamma = \frac{2\pi n_2}{\lambda A_{eff}} \qquad (2.33)$$

is known as the fiber nonlinear coefficient. The total nonlinear phase shift due to the Kerr effect after the distance L is given by

$$\varphi_{NL} = \int_0^L [\beta - \beta_0] dz \qquad (2.34)$$

Replacing Eq. (2.32) in Eq. (2.33), using Eq. (2.34) and noticing that

$$P(z) = P_0 \exp(-\alpha z) \qquad (2.35)$$

where P_0 is the launch power, and α is loss coefficient, we obtain [7].

$$\varphi_{NL} = \gamma P_0 \int_0^L \exp(-\alpha z) \, dz = \gamma P_0 \frac{1-\exp(-\alpha L)}{\alpha} = \frac{L_{eff}}{L_{NL}} \qquad (2.36)$$

where

$$L_{eff} = \frac{1-\exp(-\alpha L)}{\alpha} \qquad (2.37)$$

is the effective length, and

$$L_{NL} = \frac{1}{\gamma P_0} \tag{2.38}$$

is the nonlinear length. Physically, the nonlinear length, L_{NL}, signifies the length where the nonlinear phase change reaches 1 radian, and it possesses a length scale that the nonlinear effects become relevant for optical fibers. It can be seen from Eq. (2.36) the fiber nonlinear effects enhance when L_{NL} decreases, or equivalently power P_0 increases.

There are three kinds of fiber nonlinearities because of the Kerr effect namely SPM, XPM, and FWM. The SPM refers back to the self-induced power-dependent phase shift experienced by an optical field throughout its propagation within the optical fiber, which is accountable for spectral broadening from the optical pulses which can clearly be viewed in Figure 2.11. Figure 2.11 shows the spectral broadening of incident light into the fiber in the existence of SPM effect. Getting together with fiber dispersion, the SPM can cause temporal pulse broadening (in normal dispersion regime ($\beta_2 > 0$), or pulse compression (in anomalous dispersion regime ($\beta_2 < 0$) [51, 52].

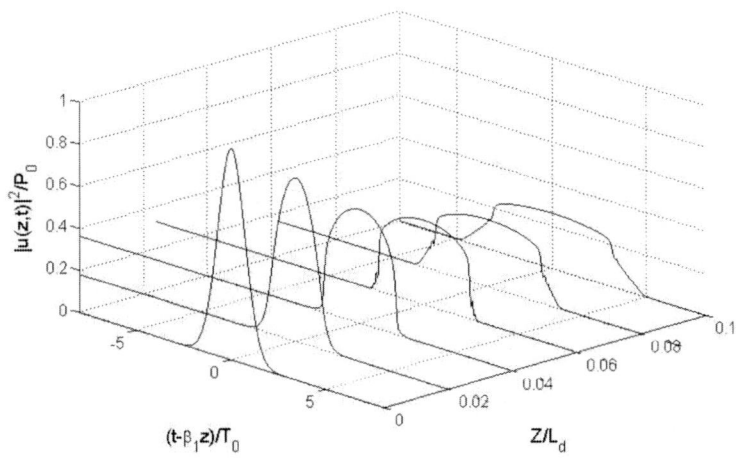

Figure 2.11. Effect of SPM on launched signal into optical fiber.

The XPM effects are extremely important for OFDM and WDM lightwave solutions because the phase of each and single optical channel is affected by the two average power along with the bit pattern of the other channels. By way of example, in WDM systems, the nonlinear phase shift of K-th channel can write as

$$\varphi_{NL} = \gamma L_{eff} P_0^{(k)} + 2 \sum_{h=1, h \neq k}^{N} \gamma L_{eff} P_0^{(k)} \qquad (2.39)$$

where $P_0^{(k)}$ signifies the peak power in K-th channel. The first term is the SPM and the second term denotes the contribution of XPM. In deriving Eq. (2.39), $P_0^{(k)}$ was assumed to be constant. In practice, time dependence of P_0 makes φ_{NL} was assumed to be constant. Refer to Eq. (2.39), the XPM caused phase change is 2 times of SPM once the optical energy of all the channels are equal. XPM causes asymmetric spectral broadening of optical pulses, timing jitter and amplitude distortion in time domain.

The FWM is the other effect that creates new frequency components. For multiuser transmission systems, for example, WDM and OFDM, with a carrier frequency of f_i, f_j and f_k the signal at new frequency $f_h = f_i + f_j - f_k$ may be produced by FWM, which leads to serious distortion when the newly generate frequency components fall into other WDM or OFDM channels. As far as transmission on fiber is concerned the nonlinear effects nearly always undesirable. After attenuation and dispersion, they provide the next major limitation on optical transmission. Indeed in some situations, they are more significant than either attenuation or dispersion. A lot of research goes into suppressing the impairments induced by SPM, XPM, and FWM for the long haul and wideband fiber optic communication systems.

2.4. Nonlinear Phase Noise Effects

A standard optical fiber includes central glass core surrounded by a cladding layer with refractive index n_c that is slightly lower than the core index n_1. For an understanding of the development of the optical industry in the optical fiber content, it is necessary to take into account wave theory based on Maxwell's equation in dispersive nonlinear media. The evolution of optical field at the transmission distance z may be described by the Nonlinear Schrodinger Equation (NLSE) [53].

For our discussion of phase noise of incident light, first, we need to consider characteristic of the medium to drive the relative wave equation. It can be assumed that the medium is isotropic which means each atomic dipole response has the same phased array less than certain condition referred to as the phase-matching condition [54]. In this case, induced polarization $P(r,t)$ is along with the direction of stimulating electrical field E (r,t) (Shieh, et. al., 2010). Additionally, there are several simplifying approximations are often made to compute the wave formula such as infinite plane-wave and slowly varying amplitude [55, 56]. To go on to drive the fiber optic wave equation, we need to review Maxwell's equations:

$$\nabla \cdot \widetilde{D} = \widetilde{\rho} \qquad (2.40)$$

$$\nabla \cdot \widetilde{B} = 0 \qquad (2.41)$$

$$\nabla \times \widetilde{E} = -\frac{\partial \widetilde{B}}{\partial t} \qquad (2.42)$$

$$\nabla \times \widetilde{H} = \frac{\partial \widetilde{D}}{\partial t} + \widetilde{J} \qquad (2.43)$$

Because that there are no free charges, currents through the silica fiber, we reach to below equations [8, 10]:

$$\tilde{\rho} = 0 \tag{2.44}$$

$$\tilde{J} = 0 \tag{2.45}$$

It is also assumed that the material is nonmagnetic such as silica (induced magnetic density M = 0) [10], so that

$$\widetilde{B} = \mu_0 \widetilde{H} \tag{2.46}$$

Electrical flux density connected with silica soluble fiber is related to polarization vector $\tilde{P}$ within nonlinear approach by means of under connection;

$$\widetilde{D} = \varepsilon_0 \widetilde{E} + \tilde{P} \tag{2.47}$$

To drive the optical wave equation, it is needed to take the curl of Eq. (2.42) and using the Eqs. (2.43), (2.45) and (2.46) to replace $\nabla \times \widetilde{B}$ by $\mu_0 (\partial \widetilde{D} / \partial t)$, Eq. (2.48) would be resultant:

$$\nabla \times \nabla \times \widetilde{E} + \mu_0 \frac{\partial^2}{\partial t^2} \widetilde{D} = 0 \tag{2.48}$$

By considering the vector analysis, the first term on the left-hand side of Eq. (2.48) can be converted to:

$$\nabla \times \nabla \times \widetilde{E} = \nabla (\nabla \cdot \widetilde{E}) - \nabla^2 \widetilde{E} \tag{2.49}$$

For simplifying the Eq. (2.48), the first term on the right-hand side of Eq. (2.49) can be negligible and wave equation can be written as Boyd, (2008):

$$\nabla^2 \widetilde{E} - \mu_0 \frac{\partial^2}{\partial t^2} \widetilde{D} = 0 \tag{2.50}$$

The electrical flux density $\tilde{D}$ and polarization vector $\tilde{P}$ can be decomposed into the linear and nonlinear part as:

$$\tilde{D} = \widetilde{D^{(1)}} + \tilde{P}^{NL} \tag{2.51}$$

$$\tilde{P} = \tilde{P}^{(1)} + \tilde{P}^{NL} \tag{2.52}$$

where linear part of $\tilde{D}$ can be expressed as

$$D^{(1)} = \varepsilon_0 \tilde{E} + \tilde{P}^{(1)} \tag{2.53}$$

By replacing Eq. (2.51) and Eq. (2.53) into Eq. (2.50) and considering $\mu_0 = 1/\varepsilon_0 c^2$ and $\varepsilon^{(1)} = n_0^2$, we obtain the expression

$$\nabla^2 \tilde{E} - \frac{n_0^2}{c^2} \frac{\partial^2}{\partial t^2} \tilde{E} = \frac{1}{\varepsilon_0 c^2} \frac{\partial^2}{\partial t^2} \tilde{P}^{NL} \tag{2.54}$$

We now assume that the light propagates along a symmetry axis, z, of the medium, and then the wave equation takes the form

$$\frac{\partial^2}{\partial z^2} \tilde{E} - \frac{n_0^2}{c^2} \frac{\partial^2}{\partial t^2} \tilde{E} = \frac{1}{\varepsilon_0 c^2} \frac{\partial^2}{\partial t^2} \tilde{P}^{NL} \tag{2.55}$$

Due to the fact that the lowest nonlinear effects originate from the third-order susceptibility $\chi^{(3)}$ in SiO_2, which is a symmetric molecule, and second order is equal to zero in this atomic structure then we can reach to $\tilde{P}^{NL}$ as [57].

$$\tilde{P}^{NL} = \varepsilon_0 \iiint_{-\infty}^{+\infty} \chi^{(3)} (t - t_1, t - t_2, t - t_3) \vdots \times$$
$$E(z, t_1)E(z, t_2)E(z, t_3)dt_1 dt_2 dt_3. \tag{2.56}$$

According to Shen (1994), $\tilde{P}^{NL}$ can approximate to

$$\tilde{P}^{NL} = \frac{3}{4}\varepsilon_0\,\chi^{(3)}\left|\tilde{E}\right|^2\tilde{E} \tag{2.57}$$

By considering Eq. (4.27) and $\tilde{E}(z,t) = u(z,t)e^{i\,(kz-\omega_0 t)}$ and applying slowly varying amplitude, right-hand side of Eq. (4.25) can be

$$\frac{\partial^2 u}{\partial z^2}e^{i(kz-\omega_0 t)} + 2ik\frac{\partial u}{\partial z}e^{i(kz-\omega_0 t)} - \frac{n_0^2}{c^2}\frac{\partial^2 u}{\partial t^2}e^{i(kz-\omega_0 t)}$$

$$+ 2i\omega_0\frac{n_0^2}{c^2}\frac{\partial u}{\partial t}e^{i(kz-\omega_0 t)} + \left(\frac{n_0^2}{c^2}k^2 - k^2\right)ue^{i(kz-\omega_0 t)}$$

$$\approx 2ik\frac{\partial u}{\partial z}e^{i(kz-\omega_0 t)} -$$

$$\frac{n_0^2}{c^2}\frac{\partial^2 u}{\partial t^2}e^{i(kz-\omega_0 t)} + 2i\omega_0\frac{n_0^2}{c^2}\frac{\partial u}{\partial t}e^{i(kz-\omega_0 t)} + \left(\frac{n_0^2}{c^2}k^2 - k^2\right)ue^{i(kz-\omega_0 t)} \tag{2.58}$$

And the left-hand side of Eq. (2.55) also can approximate to $-\frac{3}{4}\frac{\chi^{(3)}\omega_0}{k_0 c^2}|u|^2 u\,e^{i\,(kz-\omega_0 t)}$, this expression obtain

$$\frac{\partial u}{\partial z} - \frac{n_0^2}{c^2}\frac{1}{2ik}\frac{\partial^2 u}{\partial t^2} + \frac{\omega_0}{k}\frac{n_0^2}{c^2}\frac{\partial u}{\partial t} + \left(\frac{n_0^2}{c^2}k^2 - \omega_0^2\right)u = -\frac{3}{8}\frac{\chi^{(3)}\omega_0}{ik_0 c^2}|u|^2 u \tag{2.59}$$

Above mentioned equation is far more rigorous equally regarding pulse propagation over the fiber, but in the case of a weakly dispersive scenario, aforementioned expression can be simplified by ignoring any time derivative. $\partial/\partial t, \partial^2/\partial t^2$ Eq. (2.59) and yielding [58]

$$\frac{\partial u}{\partial z} = \frac{3}{8}\frac{i\chi^{(3)}\omega_0}{k_0 c^2}|u|^2 u \tag{2.60}$$

By substituting the u=v exp $(i\phi_{NL})$ in Eq. (2.60)

$$\frac{\partial v}{\partial z} = 0 \tag{2.61}$$

$$\frac{\partial \phi_{NL}}{\partial z} = \frac{3}{8} \frac{\chi^{(3)} \omega_0}{k_0 c^2} |v|^2 \tag{2.62}$$

We can obtain the total nonlinear phase shift of launching signal account for SPM and XPM with the fiber length of L as

$$\phi_{NL}(L, t) = \phi_0 + \frac{3}{8} \frac{\chi^{(3)} \omega_0}{k_0 c^2} |v|^2 Z_{eff} \tag{2.63}$$

The mentioned expression shows that the amplitude of pulse would be unaffected, while SPM gives rise to phase shape. In above equation, Z_{eff} can be defined as

$$Z_{eff} = \int_0^z e^{-\alpha \acute{z}} \, d\acute{z} = \frac{1 - e^{-\alpha z}}{\alpha} \tag{2.64}$$

In the absence of fiber loss $\alpha = 0$, $Z_{eff} = Z$, as it clear from the Eq. (2.64). What is more, the phase equation can be integrated to obtain the general solution:

$$u(z, t) = u(0, t) \exp(j\phi_{NL} t) \tag{2.65}$$

REFERENCES

[1]   D. Hillerkuss, R. Schmogrow, T. Schellinger, M. Jordan, M. Winter, G. Huber, T. Vallaitis, R. Bonk, P. Kleinow & F. Frey, (2011) "26 Tbit s-1 line-rate super-channel transmission utilizing all-optical fast Fourier transform processing," *Nature Photonics*, 5(6), 364-371.

[2]   Iraj Sadegh Amiri, Abdolkarim Afroozeh & Harith Ahmad, *Integrated micro-ring photonics: Principles and Applications as Slow light devices, Soliton generation and optical transmission.* United States: CRC Press, 2015.

[3]   SE Alavi, IS Amiri, H Ahmad, ASM Supa'at & N Fisal, (2014) "Generation and Transmission of 3× 3 W-Band MIMO-OFDM-RoF Signals Using Micro-Ring Resonators," *Applied Optics*, 53(34), 8049-8054.

[4]   S. Amiri, A. Shahidinejad, A. Nikoukar, M. Ranjbar, J. Ali & P. P. Yupapin, (2012) "Digital Binary Codes Transmission via TDMA Networks Communication System Using Dark and Bright Optical Soliton," *GSTF Journal on Computing (joc)*, 2(1), 12.

[5]   S. Amiri, S. E. Alavi, S. M. Idrus, A. S. M. Supa'at, J. Ali & P. P. Yupapin, (2014) "W-Band OFDM Transmission for Radio-over-Fiber link Using Solitonic Millimeter Wave Generated by MRR," *IEEE Journal of Quantum Electronics*, 50(8), 622 - 628.

[6]   IS Amiri, SE Alavi, N Fisal, ASM Supa'at & H Ahmad, (2014) "All-Optical Generation of Two IEEE802.11n Signals for 2×2 MIMO-RoF via MRR System," *IEEE Photonics Journal*, 6(6),

[7]   S. Amiri, M. Ebrahimi, A. H. Yazdavar, S. Gorbani, S. E. Alavi, Sevia M. Idrus & J. Ali, (2014) "Transmission of data with orthogonal frequency division multiplexing technique for communication networks using GHz frequency band soliton carrier," *IET Communications*, 8(8), 1364 – 1373.

[8]   S. Amiri, S. E. Alavi & J. Ali, (2013) "High Capacity Soliton Transmission for Indoor and Outdoor Communications Using Integrated Ring Resonators," *International Journal of Communication Systems*, 28(1), 147–160.

[9]   S. E. Alavi, I. S. Amiri, S. M. Idrus, ASM Supa'at, J. Ali & P. P. Yupapin, (2014) "All Optical OFDM Generation for IEEE802.11a Based on Soliton Carriers Using MicroRing Resonators," *IEEE Photonics Journal*, 6(1),

[10] S. Amiri, S. E. Alavi, H. Ahmad, A. S. M. Supa'at & N. Fisal, (2014) "Numerical Computation of Solitonic Pulse Generation for Terabit/Sec Data Transmission," *Optical and Quantum Electronics*,

[11] S. E. Alavi, I. S. Amiri, S. M. Idrus & A. S. M. Supa'at, (2014) "Generation and Wired/Wireless Transmission of IEEE802.16m Signal Using Solitons Generated By Microring Resonator," *Optical and Quantum Electronics*,

[12] S. E. Alavi, I. S. Amiri, M. Khalily, A. S. M. Supa' at, N. Fisal, H. Ahmad & S. M. Idrus, (2014) "W-Band OFDM for Radio-over-Fibre Direct-Detection Link Enabled By Frequency Nonupling Optical Up-Conversion," *IEEE Photonics Journal* 6(6),

[13] Iraj Sadegh Amiri, Sayed Ehsan Alavi & Sevia Mahdaliza Idrus, *Soliton Coding for Secured Optical Communication Link*. USA: Springer, 2014.

[14] M. E. Marhic, (1987) "Discrete Fourier transforms by single-mode star networks," *Optics letters*, 12(1), 63-65.

[15] K. B. Howell, *Principles of Fourier Analysis*: CRC Press, 2001.

[16] Siegman, (2001) "Fiber Fourier optics," *Optics letters*, 26(16), 1215-1217.

[17] Siegman, (2002) "Fiber Fourier optics: previous publication," *Optics letters*, 27(6), 381-381.

[18] D. Hillerkuss, M. Winter, M. Teschke, A. Marculescu, J. Li, G. Sigurdsson, K. Worms, S. Ben Ezra, N. Narkiss & W. Freude, (2010) "Simple all-optical FFT scheme enabling Tbit/s real-time signal processing," *Optics express*, 18(9), 9324-9340.

[19] Y. S. Neo, S. M. Idrus, M. F. Rahmat, S. E. Alavi & I. S. Amiri', (2014) "Adaptive Control for Laser Transmitter Feedforward Linearization System," *IEEE Photonics Journal* 6(4),

[20] Abdolkarim Afroozeh, Iraj Sadegh Amiri, Alireza Zeinalinezhad & Harith Ahmad, *Characterization and Controlling of Soliton Signals Generated by Semiconductor Microring Resonators*. USA: Springer, 2015.

[21] Afroozeh, I. S. Amiri, A. Zeinalinezhad, S. E. Pourmand, S. E. Alavi & H. Ahmad, (2015) "Comparison of Control Light using Kramers-Kronig Method by Three Waveguides," *Journal of Computational and Theoretical Nanoscience (CTN)*, 12(8),

[22] S. Amiri, S. E. Alavi, Sevia M. Idrus, A. Nikoukar & J. Ali, (2013) "IEEE 802.15.3c WPAN Standard Using Millimeter Optical Soliton Pulse Generated By a Panda Ring Resonator," *IEEE Photonics Journal*, 5(5), 7901912.

[23] S. E. Alavi, I. S. Amiri, A. S. M. Supa'at & S. M. Idrus, (2015) "Indoor Data Transmission Over Ubiquitous Infrastructure of Powerline Cables and LED Lighting," *Journal of Computational and Theoretical Nanoscience (CTN)*, 12(4),

[24] Iraj Sadegh Amiri & Abdolkarim Afroozeh, *Ring Resonator Systems to Perform the Optical Communication Enhancement Using Soliton*. USA: Springer, 2014.

[25] Iraj Sadegh Amiri, Ali Nikoukar & Sayed Ehsan Alavi, *Soliton and Radio over Fiber (RoF) Applications. Saarbrücken,* Germany: LAP LAMBERT Academic Publishing, 2014.

[26] Afroozeh, I. S. Amiri, M. A. Jalil, M. Kouhnavard, J. Ali & P. P. Yupapin, (2011) "Multi Soliton Generation for Enhance Optical Communication," *Applied Mechanics and Materials*, 83 136-140.

[27] S. Amiri, S. Soltanmohammadi, A. Shahidinejad & j. Ali, (2013) "Optical quantum transmitter with finesse of 30 at 800-nm central wavelength using microring resonators," *Optical and Quantum Electronics*, 45(10), 1095-1105.

[28] Abdolkarim Afroozeh, Iraj Sadegh Amiri & Alireza Zeinalinezhad, *Micro Ring Resonators and Applications: New Technology of Optics*. Saarbrücken, Germany: LAP LAMBERT Academic Publishing, 2014.

[29] Iraj Sadegh Amiri, Sayed Ehsan Alavi, S. M. Idrus, Abdolkarim Afroozeh & Jalil Ali, *Soliton Generation by Ring Resonator for Optical Communication Application*. New York: Nova Science Publishers, 2014.

[30] S. Amiri, R. Ahsan, A. Shahidinejad, J. Ali & P. P. Yupapin, (2012) "Characterisation of bifurcation and chaos in silicon microring resonator," *IET Communications*, 6(16), 2671-2675.

[31] S. Amiri & J. Ali, (2014) "Generating Highly Dark–Bright Solitons by Gaussian Beam Propagation in a PANDA Ring Resonator," *Journal of Computational and Theoretical Nanoscience (CTN)*, 11(4), 1092-1099.

[32] S. Amiri, S. E. Alavi & H. Ahmad, (2015) "Analytical Treatment of the Ring Resonator Passive Systems and Bandwidth Characterization Using Directional Coupling Coefficients," *Journal of Computational and Theoretical Nanoscience (CTN)*, 12(3),

[33] I. S. Amiri & A. Afroozeh, Soliton Generation Based Optical Communication, in *Ring Resonator Systems to Perform Optical Communication Enhancement Using Soliton*, ed USA: Springer, 2014.

[34] S. Amiri, S. E. Alavi & S. M. Idrus, Introduction of Fiber Waveguide and Soliton Signals Used to Enhance the Communication Security, in *Soliton Coding for Secured Optical Communication Link*, ed USA: Springer, 2015, pp. 1-16.

[35] Shahidinejad, A. Nikoukar, I. S. Amiri, M. Ranjbar, A. Shojaei, J. Ali & P. Yupapin, (2012), "Network system engineering by controlling the chaotic signals using silicon micro ring resonator," in *Computer and Communication Engineering (ICCCE) Conference*, Malaysia, 765-769.

[36] S. Amiri, M. H. Khanmirzaei, M. Kouhnavard, P. P. Yupapin & J. Ali, Quantum Entanglement using Multi Dark Soliton Correlation for Multivariable Quantum Router, in *Quantum Entanglement* A. M. Moran, Ed., ed New York: Nova Science Publisher, 2012, pp. 111-122.

[37] S. Amiri, A. Nikoukar, A. Shahidinejad, J. Ali & P. Yupapin, (2012), "Generation of discrete frequency and wavelength for secured computer networks system using integrated ring

resonators," in *Computer and Communication Engineering (ICCCE) Conference*, Malaysia, 775-778.

[38] S. Amiri & J. Ali, (2014) "Optical Quantum Generation and Transmission of 57-61 GHz Frequency Band Using an Optical Fiber Optics," *Journal of Computational and Theoretical Nanoscience (CTN)*, 11(10), 2130-2135.

[39] S. Amiri, P. Naraei & J. Ali, (2014) "Review and Theory of Optical Soliton Generation Used to Improve the Security and High Capacity of MRR and NRR Passive Systems," *Journal of Computational and Theoretical Nanoscience (CTN)*, 11(9), 1875-1886.

[40] S. Amiri, S. E. Alavi, M. Bahadoran, A. Afroozeh & H. Ahmad, (2015) "Nanometer Bandwidth Soliton Generation and Experimental Transmission within Nonlinear Fiber Optics Using an Add-Drop Filter System," *Journal of Computational and Theoretical Nanoscience (CTN)*, 12(2),

[41] I. S. Amiri, M. R. K. Soltanian, S. E. Alavi & H. Ahmad, (2015) "Multi Wavelength Mode-lock Soliton Generation Using Fiber Laser Loop Coupled to an Add-drop Ring Resonator," *Optical and Quantum Electronics*,

[42] I. S. Amiri & A. Afroozeh, Mathematics of Soliton Transmission in Optical Fiber, in *Ring Resonator Systems to Perform Optical Communication Enhancement Using Soliton*, ed USA: Springer, 2014.

[43] S. Amiri, S. E. Alavi & S. M. Idrus', (2014) "Solitonic Pulse Generation and Characterization by Integrated Ring Resonators," presented at the *5th International Conference on Photonics 2014 (ICP2014)*, Kuala Lumpur.

[44] S. Amiri & J. Ali, (2013) "Data Signal Processing Via a Manchester Coding-Decoding Method Using Chaotic Signals Generated by a PANDA Ring Resonator," *Chinese Optics Letters*, 11(4), 041901(4).

[45] S. Amiri, A. Afroozeh & M. Bahadoran, (2011) "Simulation and Analysis of Multisoliton Generation Using a PANDA Ring Resonator System," *Chinese Physics Letters*, 28(10), 104205.

[46] S. Amiri, J. Ali & P. P. Yupapin, (2012) "Enhancement of FSR and Finesse Using Add/Drop Filter and PANDA Ring Resonator Systems," *International Journal of Modern Physics B*, 26(04), 1250034.

[47] S. Amiri, A. Afroozeh, I. N. Nawi, M. A. Jalil, A. Mohamad, J. Ali & P. P. Yupapin, (2011) "Dark Soliton Array for communication security," *Procedia Engineering*, 8 417-422.

[48] I. S. Amiri & A. Afroozeh, Integrated Ring Resonator Systems, in *Ring Resonator Systems to Perform Optical Communication Enhancement Using Soliton*, ed. USA: Springer, 2014.

[49] I. S. Amiri & A. Afroozeh, Introduction of Soliton Generation, in *Ring Resonator Systems to Perform Optical Communication Enhancement Using Soliton*, ed. USA: Springer, 2014.

[50] Sadegh Amiri, M. Nikmaram, A. Shahidinejad & J. Ali, (2013) "Generation of potential wells used for quantum codes transmission via a TDMA network communication system," *Security and Communication Networks*, 6(11), 1301-1309.

[51] Iraj Sadegh Amiri & Harith Ahmad, *Optical Soliton Communication Using Ultra-Short Pulses*. USA: Springer, 2014.

[52] S. Amiri, S. E. Alavi & S. M. Idrus, Theoretical Background of Microring Resonator Systems and Soliton Communication, in *Soliton Coding for Secured Optical Communication Link*, ed USA: Springer, 2015, pp. 17-39.

[53] G. P. Agrawal, *Nonlinear Fiber Optics*: Academic Press, 2007.

[54] R. Boyd, (2008) *Nonlinear Optics* 3rd edn (New York: Academic).

[55] R. R. Alfano, (1989) "*The supercontinuum laser source.*"

[56] W. Shieh & I. Djordjevic, *Orthogonal frequency division multiplexing for optical communications vol. 14*: Academic Press, 2010.

[57] D. Mills, *Nonlinear optics*: Springer Berlin etc., 1991.

[58] M. Nazarathy, J. Khurgin, R. Weidenfeld, Y. Meiman, P. Cho, R. Noe, I. Shpantzer & V. Karagodsky, (2008) "Phased-array cancellation of nonlinear FWM in coherent OFDM dispersive multi-span links," *Optics express*, 16(20), 15777-15810.

Chapter 3

SYSTEM CONFIGURATION OF COUPLER-BASED ALL-OPTICAL OFDM TRANSMISSION SYSTEM

ABSTRACT

In all optical systems, the transmitter side consists of comb power generator, wavelength selected switch and optical QAM generator. Comb power generator generates channels with frequency separation $\Delta f = 25$ GHz. Subsequently, Wavelength selected Switch (WSS) was used to split subcarriers and then subcarriers are modulated individually with Optical QAM modulators. In the following session, the different parts of all-optical OFDM will be explained in details.

Keywords: MZM, comb generator, QAM, model of all-optical OFDM

3.1. M-QAM SYSTEM

As it was explained in the last chapter, multilevel modulation signals like differential quadrature phase-shift keying (DQPSK) have obtained renewed attention to boost the spectral efficiency of any lightwave communication method. Increasing the bandwidth using the optical communication program, bring more awareness of research in substantial data rate transmitting system without enhancing the bandwidth like quadrature-amplitude modulation (QAM) signal modulation. In this part we will present that irrespective of the number of points, all QAM signals can be generated using a single or dual-generate MZM. The M-QAM is multilevel signaling formats; bring more spectral efficiency compared to standard NRZ and RZ modulation technique. The high spectral efficiency of formats enables increased capacity of communication systems in the limited bandwidth of the optical amplifiers. The strength penalty associated with the multilevel signaling can be mitigated by coherent discovery. As it can be seen in Figure 3.1 the M-QAM transmission part can be generated making use of four different transmitter architectures as a stick to K.P. [1, 2].

1. Pair of single drive MZMs
2. Pair of dual drive MZMs
3. Single twin drive MZM
4. Sequential phase and amplitude modulators

In multiple level modulation electronic digital transmission, m bites are collected and mapped to a sophisticated symbol chosen from an alphabet

$$d(k) = d_{rk} + jd_{jk} \in \{d_0, d_1, \dots, d_{M-1}\}, M = 2^m \qquad (3.1)$$

For each symbol interval (K) of length $T_s = m.T_B$ where $1/T_B$ is the bit rate, one of several symbols of this alphabet is assigned, based on the respective bit mixture and defined in the so called constellation diagram. Different from direct detection systems, in coherent optical systems the constellation points can be equally spaced as for electrical systems, because the noise does not depend on the power of the detected symbols. For further calculations it is useful to scale d_{rk} and d_{ik} to unity:

$$S(k)= i_k + jq_k \, , i_k = \frac{d_{rk}}{\max\{d_{rk}\}} \, , q_k = \frac{d_{ik}}{\max\{d_{ik}\}} \tag{3.2}$$

The complex envelope, as well as its components, is understood to be

$$A(t) = a(t).\,e^{j\varphi(t)} = \frac{1}{\sqrt{2}} s(k).\,p(t - KT_s) = \frac{1}{\sqrt{2}} I(t) + j\frac{1}{\sqrt{2}} Q(t) \tag{3.3a}$$

$$I(t) = \Sigma_k(i_k.\,p(t - KT_s)), Q(t) = \Sigma_k(q_k \cdot p(t - KT_s)) \tag{3.3b}$$

$$a(t) = \frac{1}{\sqrt{2}}(t)\sqrt{I^2(t) + Q^2(t)}, \varphi(t)=\arctan\frac{Q(t)}{I(t)} \tag{3.3c}$$

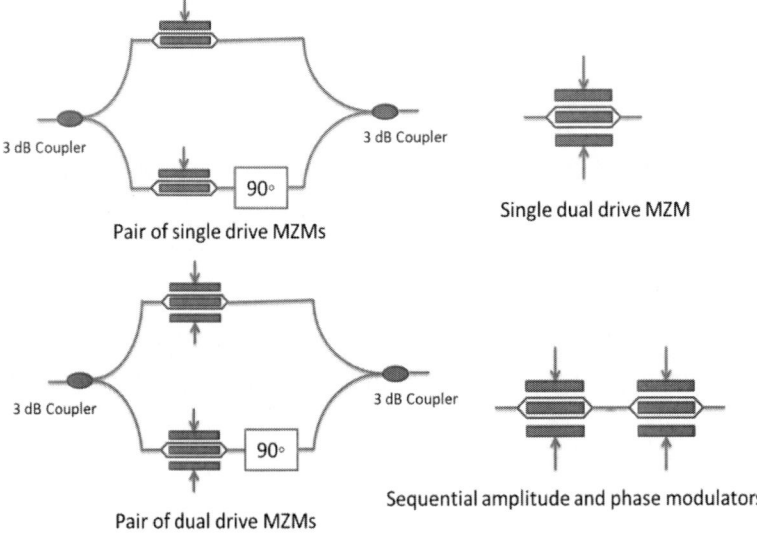

Figure 3.1. M-QAM Schematics.

A(t) and φ(t) are symbolized the complex envelope's amplitude and stage respectively. It has to be noticed that the arc tan function is defined for all four quadrants by case differentiae. I(t) and Q(t) describe the in phase part and the quadrature element of the complex envelope, respectively. p(t) is the pulse model of the electrical traveling signals.

M-QAM modulation can also be called as multi-structure due to the optical portion of the transmitters is common. To appreciate a special modulation structure, the electrical driving has to be adapted. An optical multi-level modulation signal might be generated either from an amplitude modulator followed by a phase modulator or from IQ modulator comprised of two arms with two orthogonal carriers, where the phase component of the complex envelope modulate the optical carrier within the I-left arm and quadrature part modulates the 90°stage shifted optical carrier in the Q left arm. Amplitude modulation needs to perform in each arm and the electric driving signals use a smaller number of states. A further reduction of the volume of states can be arrived at by replacing the amplitude modulation in each arm by separate phase and intensity modulations. Then an intensity modulator is driven by a unipolar RF driving signal at a DC bias point at-π and also the phase modulator adjustments the phase between 0 and π for positive and negative values of i_k and q_k respectively. Furthermore, it is possible to realize the phase and amplitude modulation with just one component, using a two drive MZM. The RF driving signals for phase and intensity modulation really need to be electrically combined prior to inject into MZM inputs. In the subsequent it is, shown, that this differences in the transmitter built-ups and different characteristics of driving signals bring about different properties of the multi-levels modulation signals.

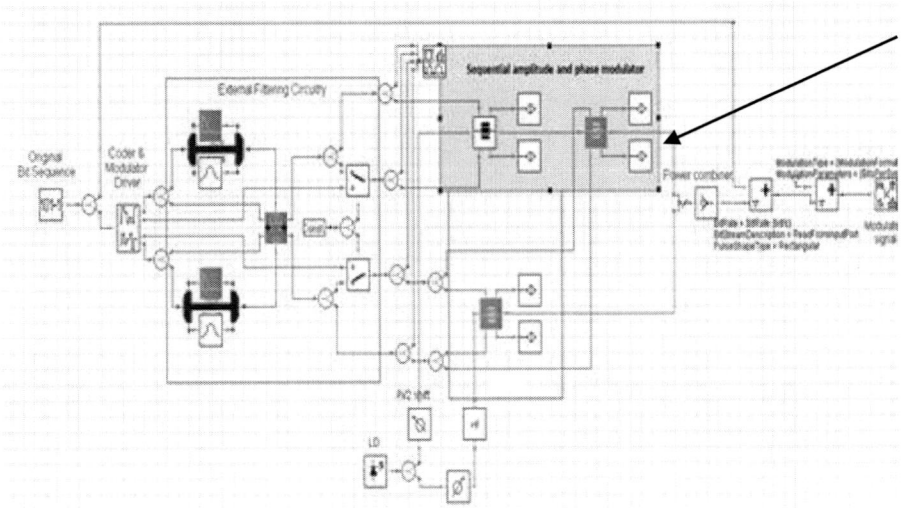

Figure 3.2. a) M-QAM sequential and phase modulator.

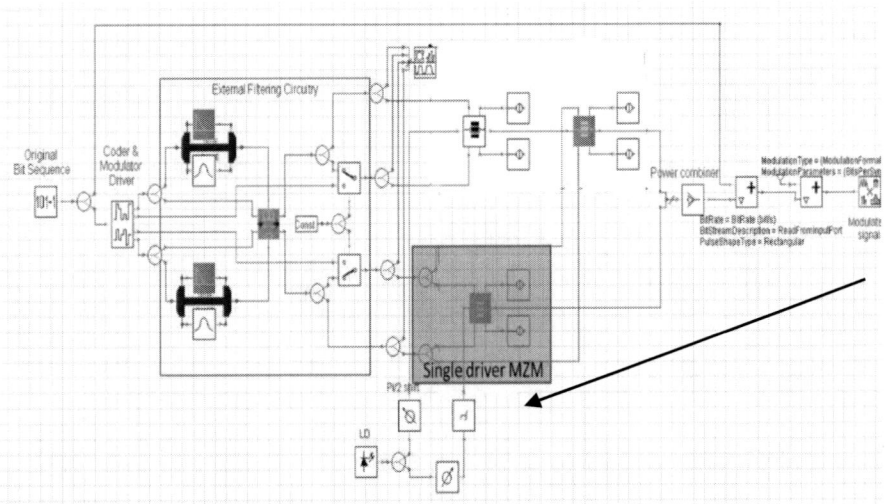

Figure 3.2. b) M-QAM Single driver MZM Configuration.

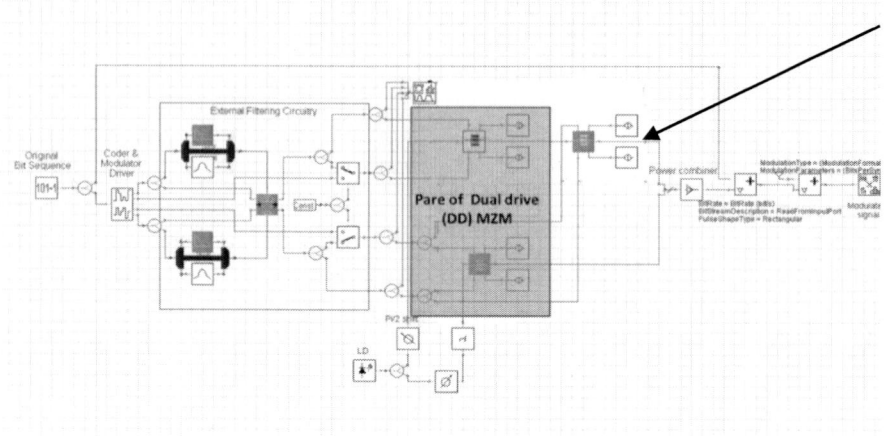

Figure 3.2. c) M-QAM Pair of Dual drive(DD) MZM configuration.

Table 3.1. Expressions for the electrical driving signals for different transmitters

Configuration	Driving signals
Sequential amplitude and phase modulator	$V_{RF}^{AM}(t) = \dfrac{2V_{\pi,rf}}{\pi} \arcsin(\dfrac{\sqrt{i_K^2 + q_K^2}}{\sqrt{2}})$
	$V_{RF}^{PM}(t) = \dfrac{2V_{\pi,rf}}{\pi} \arctan(\dfrac{q_k}{i_k})$
Single dual driver MZM	$V_{RF}^{I}(t) = \dfrac{2V_{\pi,rf}}{\pi} \cdot \sum_k \arcsin(i_k)\, p(t - kT_s)$
Pair of single drive MZM	$V_{RF}^{IM,I}(t) = \dfrac{2V_{\pi,RF}}{\pi} \cdot \sum_k (\arcsin(|i_k|)..\, P_{IM}(t - kT_s))$
	$V_{RF}^{PM,I}(t) = \dfrac{V_{\pi,RF}}{2} \cdot \sum_k ((-\text{sign}(i_k) + 1).\, P_{PM}(t - kT_s))$
Pair of Dual drive (DD) MZM	$V_{RF}^{UP,I}(t) = \dfrac{1}{2} V_{RF}^{IM,I}(t) + V_{RF}^{PM,,I}(t)$ $V_{RF}^{low,I}(t) = -\dfrac{1}{2} V_{RF}^{IM,I} + V_{RF}^{PM,I}(t)$

In order to see the difference of QAM modulation configuration, the Constellation diagram of different configuration is demonstrated in Figure 3.3. A constellation diagram is a representation of a signal modulated by a digital modulation scheme such as quadrature amplitude modulation.

It displays the phase and amplitude data as a two-dimensional diagram in the real and imaginary plane at all of the symbol sampling time. The diagrams shown there is insignificant difference in QAM modulation configuration. Due to use of dual drive (DD) MZM in conventional QAM systems, the rest of our works we set the Pair of Dual drive (DD) MZM as our configuration.

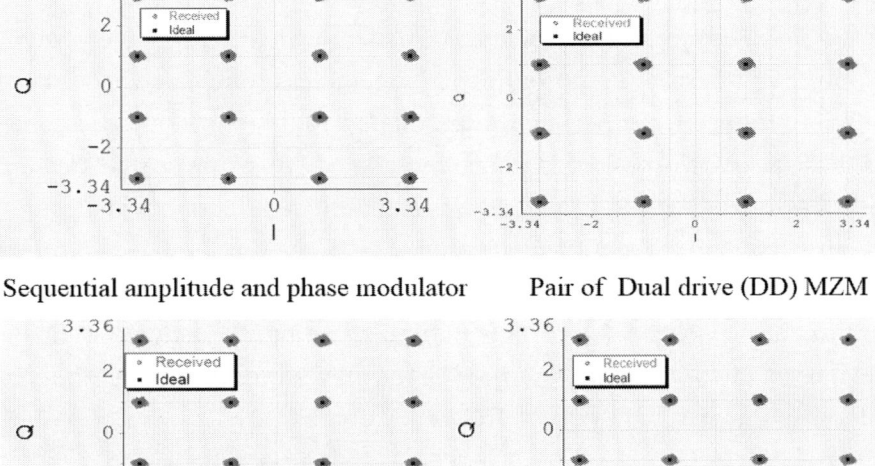

Sequential amplitude and phase modulator   Pair of Dual drive (DD) MZM

Pair of single drive MZM   Single dual driver MZM

Figure 3.3. Constellation diagram of different configuration.

3.2. OPTICAL COMB GENERATOR

Optical comb generator offers important results within all-optical transmission systems [3-10]. At present, mode-locked lasers in addition to stage modulation tend to be two processes to produce optical comb [11]. Flat-top optical frequency comb produced via stage modulation from the CW laser beam offers advantages of the rate of recurrence spacing adjustability, therefore, it provides drawn scientists curiosity for quite some time. Time-to-frequency (TTF) is one method of flat-top optical frequency comb produced via stage modulation from the CW laser beam. The quantity of narrow comb outlines is an immediate percentage to be able to stage modulation catalog this way. When the half-wave related to stage voltage modulator is set, we are able to increase the amplitude associated with the sinusoidal waveform to acquire extra narrow comb outlines. Nonetheless, one stage modulator can't be produced too big amplitude related to a sinusoidal waveform. So the use of two cascaded stage modulators to enhance the flat-top optical frequency comb is suggested. The perfect parabolic waveform is actually difficult to get, whilst sinusoidal waveform enables you to substitute perfect parabolic waveform. The particular plan is actually confirmed in Figure 3.4 with this particular set up, the actual nonlinear effect associated with strength modulator may be used to boost the real flatness associated with optical frequency combs.

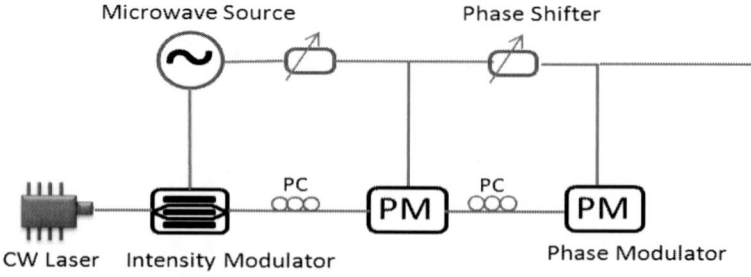

Figure 3.4. Comb generator schematic model.

The flatness of optical frequency comb changes with characteristics of the optical pulse and optical pulse is related to the amplitude of sine signal Γ_m and DC bias Γ_B applied in intensity modulator.

Where:

$$\Gamma_m = \frac{V_a}{V_\pi}, and\ \Gamma_B = \frac{V_{DC}}{V_\pi}, \quad V_\pi \text{ is half-wave voltage intensity}$$

modulator (V_π=6.5 v).

The half-wave voltage of both phase modulators is 3v.

Frequency of microwave is space channel =25GHz

$$\Gamma_m = \frac{V_a}{6.5} = 0.5 \times \pi \Rightarrow V_a = 3.25 \times \pi \tag{3.4}$$

$$\Gamma_B = \frac{V_{DC}}{6.5} = 0.35 \times \pi \Rightarrow V_{DC} = 2.2275 \times \pi \tag{3.5}$$

$$E_{out} = E_{int}.\exp\left(i \times \Delta\theta_1 \times \cos(\omega.t + \Delta\varphi_1) + \left(i \times \Delta\theta_1 \times \cos(\omega.t + \Delta\varphi_1)\right)\right) \tag{3.6}$$

$\Delta\theta_1$ and $\Delta\theta_2$ are the phase modulator indices.

$\Delta\varphi_1$ and $\Delta\varphi_2$ are the phase shifts of sine signals applied in intensity and phase modulators.

With these values can obtain flattop comb. But there are other combinations as well.

$$E_{out} = E_{int}.\exp\left(i \times 16 \times \cos(\omega.t + 0)\right) + \left(i \times 16 \times \cos(\omega.t + 0)\right)$$

$$(3.7)$$

$$E_{int} = \cos(197.3THz \times t) \text{ Which is CW Laser} \qquad (3.8)$$

3.3. THE ALL-OPTICAL OFDM TRANSMITTER

The OFDM transmission system set-up is shown in Figure 3.5. In the transmitter, comb generator, which generates channels with frequency separation $\Delta f = 25$ GHz. Subsequently, Wavelength Selective Switch (WSS) is used to split subcarriers to odd and evens and then subcarriers are modulated individually with Optical QAM modulators. The proposed scheme of Optical QAM is proposed in last section [12]. As the Figure 3.5 shows, after QAM modulators, the data combine together using 3d couplers to form the OFDM signal [13, 14]. To preserve the orthogonality of OFDM, the symbol duration of OFDM is required to be $T_s = \frac{1}{\Delta f}$. About the bandwidth limitations of the modulator and the resulting finite rise and fall times inside OFDM symbol duration T_s, the symbol undergoes distortion in amplitude and phase and this event leads to crosstalk. In order to mitigate mentioned distortion, we need to implement Cyclic Prefix (CP) of 20% to shift the rise and fall times of the transmitter and receiver out of the symbol duration window [15].

In addition, for the transmission line, in this book, the Standard Single Mode Fiber (SSMF) is used with length L = 80 km, Fiber attenuation factor α=0.2 dB/km, nonlinear refractive index n_2=2.5 × 10^{-20}m²/w, fiber core effective index A_{eff}=80μm^2. As it clear from the literatures, due to soliton effects, the positive chirp effect from Phase noise compensate by positive chirp effect from dispersion. In order to measure the real phase noise in system, fiber dispersion coefficient set at zero.

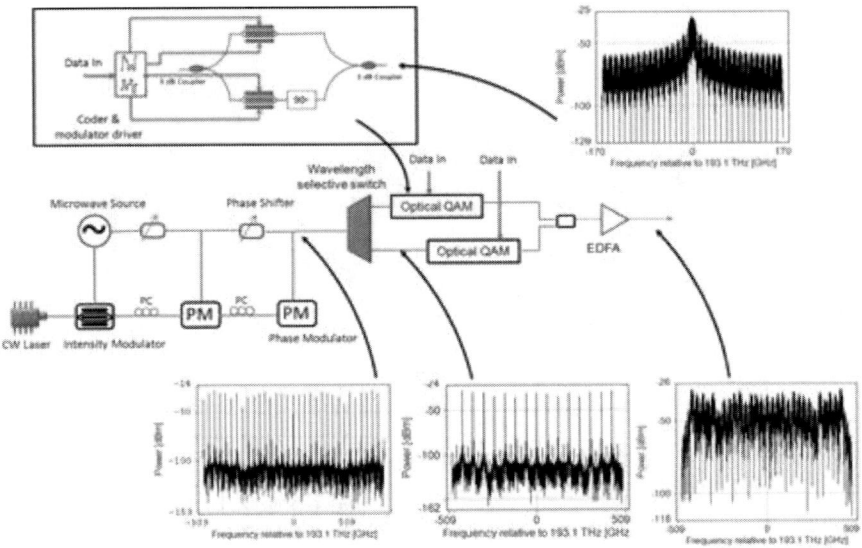

Figure 3.5. Schematic diagram of an all-optical OFDM transmission system with comb generator and all optical QAM modulation.

3.4. THE ALL-OPTICAL OFDM RECEIVER

On the receiver side, as shown in Figure 3.6, the received OFDM signal is processed using the low-complexity all-optical FFT (OFFT) circuit. This scheme of N-order OFFT is used to perform both serial-to-parallel conversions and FFT in the optical domain using N-1 cascade of Mach-Zehnder Interferometers (MZIs) with subsequent time gates, optical phase modulators, and $C_{MZI} = 2(N - 1)$ couplers. For the four orders OFFT, as shown in Figure 3.6, the three numbers of MZIs is required. The first MZI time delay is adjusted to $\frac{T_s}{2}$, and the time delay of two other subsequent parallel MZIs are set to $\frac{T_s}{4}$. After being processed by OFFT, the resulting signals are sampled by electro-absorption modulators (EAM). Afterwards, the output from EAM is fed to an optical fourth-order super-Gaussian Bandpass filter and detected using QAM demodulators. Bit error rate of resulting signals was measured using a BER.

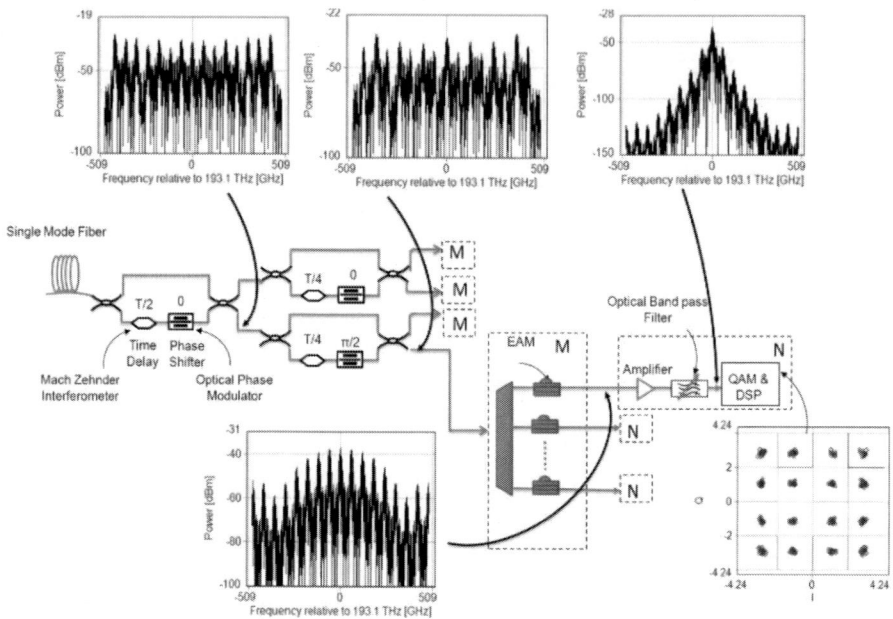

Figure 3.6. The receiver components of all-optical transmission system with all-optical FFT schemes.

3.5. TRANSMITTING MODEL OF ALL-OPTICAL OFDM

The above is the general picture for wave propagation through the fiber, but the aim of these scientific studies is to survey the effect of phase noise for All-optical OFDM, then the analytical model of the system will need to take into account in previously mentioned simulation [16]. The proposed block diagram of all-optical OFDM is demonstrated in Figure 3.7.

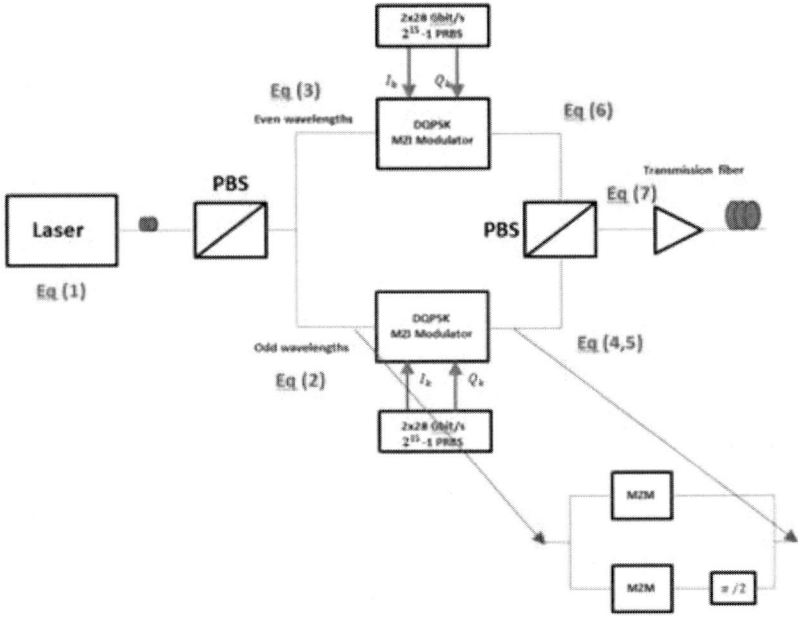

Figure 3.7. Block diagram of all-optical OFDM.

The optical output signal of the serial Square QAM transmitter for NRZ pulse shape is given by

$$E_s(t) = laser.\cos\left(\frac{U_{IM}(t)}{2V_{pi}}pi\right).e^{j\frac{U_{PM}(t)}{V_{pi}}pi} \tag{3.9}$$

In order to adjust the desired amplitude and phase levels, the multi-level electrical driving signals for the MZM and the PM must be chosen as

$$U_{IM}(t) = -V_{pi} + 2\frac{V_{pi}}{pi}.\sum_k^D(\arcsin\left(\frac{\sqrt{I_k^2+Q_k^2}}{\sqrt{2}}\right).p(t - kT_s), k =$$

$$1,2,3,\dots D \tag{3.10}$$

$$U_{PM}(t) = \frac{V_{pi}}{pi} \cdot \sum_{k}^{D}(arg[I_k, Q_k] \cdot p(t - kT_s), k = 1,2,3, \ldots D \quad (3.11)$$

where D is number of bits per samples (bits), I_k and Q_k are the pre-coded QAM-4 input data streams.

The laser in optical communication system can be modeled by $Laser = a. \text{Re}[\exp(i2\pi f_{laser}t)]$. Where (a) is the amplitude of the pumped laser and f_{laser} is the frequency of the laser at 197.3 THz [17]. So the odd comb lines can be formulated by:

$$oddcombs = \sum_{n=2k+1}^{N/2} Laser \times \exp(i2\pi f_{spacing}t.n) \, k = 1,2,3, \ldots$$
$$(3.12)$$

where (N) is the number of comb lines, $f_{spacing}$ is the channel spacing between the comb lines and n is index number of odd comb lines.

With the same method, even comb lines can be formulated easily.

$$evencombs = \sum_{m=2k}^{N/2} Laser \times \exp(i2\pi f_{spacing}t.m) \, k = 1,2,3, \ldots$$
$$(3.13)$$

The summation of above equations will be the transmitting modulated signal as follow:

$$u_i(0,t) = e^{i2\pi f_{laser}t}. \sum_{n=k}^{N} e^{i2\pi f_{spacing}t.n}. \left(\sum_{k=1}^{D} s_i(k)\right) k = 1,2,3, \ldots, D$$
$$(3.14)$$

The analytical output results of the all-optical OFDM are depicted in Figure 3.8. The output of Spectrum of comb generator follow by All-optical OFDM signal is depicted in Figure 3.8(a) and Figure 3.8(b) respectively.

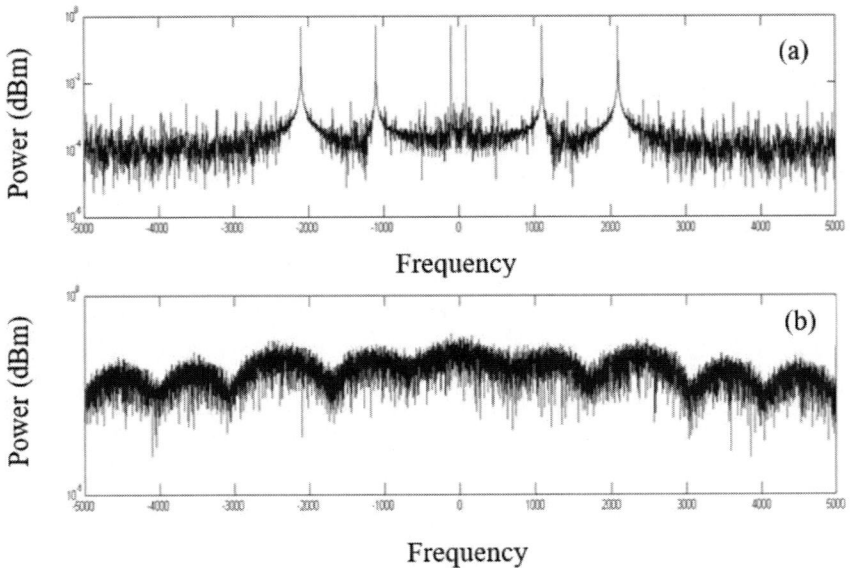

Figure 3.8. Analytical results of the all-optical OFDM.

REFERENCES

[1] K. P. Ho & H. W. Cuei, (2005) "Generation of arbitrary quadrature signals using one dual-drive modulator," *Journal of lightwave technology*, 23(2), 764.

[2] M. Seimetz, (2005), "Multi-format transmitters for coherent optical M-PSK and M-QAM transmission," in *Transparent Optical Networks, 2005, Proceedings of 2005 7th International Conference* 225-229.

[3] Abdolkarim Afroozeh, Iraj Sadegh Amiri, Alireza Zeinalinezhad & Harith Ahmad, *Characterization and Controlling of Soliton Signals Generated by Semiconductor Microring Resonators.* USA: Springer, 2015.

[4] S. Amiri, S. E. Alavi, Sevia M. Idrus, A. Nikoukar & J. Ali, (2013) "IEEE 802.15.3c WPAN Standard Using Millimeter

Optical Soliton Pulse Generated By a Panda Ring Resonator," *IEEE Photonics Journal*, 5(5), 7901912.

[5]   I. S. Amiri, S. E. Alavi, N. Fisal, A. S. M. Supa'at & H Ahmad, (2014) "All-Optical Generation of Two IEEE802.11n Signals for 2×2 MIMO-RoF via MRR System", *IEEE Photonics Journal*, 6(6),

[6]   S. E. Alavi, I. S. Amiri, M. Khalily, A. S. M. Supa' at, N. Fisal, H. Ahmad & S. M. Idrus, (2014) "W-Band OFDM for Radio-over-Fibre Direct-Detection Link Enabled By Frequency Nonupling Optical Up-Conversion," *IEEE Photonics Journal* 6(6),

[7]   S. E. Alavi, I. S. Amiri, S. M. Idrus, ASM Supa'at, J. Ali & P. P. Yupapin, (2014) "All Optical OFDM Generation for IEEE802.11a Based on Soliton Carriers Using MicroRing Resonators," *IEEE Photonics Journal*, 6(1),

[8]   Y. S. Neo, S. M. Idrus, M. F. Rahmat, S. E. Alavi & I. S. Amiri', (2014) "Adaptive Control for Laser Transmitter Feedforward Linearization System," *IEEE Photonics Journal* 6(4),

[9]   Iraj Sadegh Amiri & Abdolkarim Afroozeh, *Ring Resonator Systems to Perform the Optical Communication Enhancement Using Soliton.* USA: Springer, 2014.

[10] S. Amiri, M. Ebrahimi, A. H. Yazdavar, S. Gorbani, S. E. Alavi, Sevia M. Idrus & J. Ali, (2014) "Transmission of data with orthogonal frequency division multiplexing technique for communication networks using GHz frequency band soliton carrier," *IET Communications*, 8(8), 1364 – 1373.

[11] I. S. Amiri, M. R. K. Soltanian, S. E. Alavi & H. Ahmad, (2015) "Multi Wavelength Mode-lock Soliton Generation Using Fiber Laser Loop Coupled to an Add-drop Ring Resonator," *Optical and Quantum Electronics*,

[12] R. Griffin & A. Carter, (2002), "Optical differential quadrature phase-shift key (oDQPSK) for high capacity optical transmission," in *Optical Fiber Communication Conference and Exhibit, 2002. OFC 2002* 367-368.

[13] R. Schmogrow, D. Hillerkuss, M. Dreschmann, M. Huebner, M. Winter, J. Meyer, B. Nebendahl, C. Koos, J. Becker & W. Freude, (2010) "Real-time software-defined multiformat transmitter generating 64QAM at 28 GBd," *Photonics Technology Letters, IEEE*, 22(21), 1601-1603.

[14] D. Hillerkuss, M. Winter, M. Teschke, A. Marculescu, J. Li, G. Sigurdsson, K. Worms, S. Ben Ezra, N. Narkiss & W. Freude, (2010) "Simple all-optical FFT scheme enabling Tbit/s real-time signal processing," *Optics express*, 18(9), 9324-9340.

[15] Q. Yang, S. Chen, Y. Ma & W. Shieh, (2009) "Real-time reception of multi-gigabit coherent optical OFDM signals," *Optics express*, 17(10), 7985-7992.

[16] S. Kumar, *Impact of Nonlinearities on Fiber Optic Communications vol. 7*: Springer Science & Business Media, 2011.

[17] M. Seimetz, *High-order modulation for optical fiber transmission vol. 143*: Springer Science & Business Media, 2009.

Chapter 4

RESULTS AND DISCUSSIONS OF PHASE NOISE EFFECTS BASED ON FIBER LENGTH, CHANNEL NUMBERS AND LASER POWER

ABSTRACT

The effect of phase noise on both the 4-QAM and 16-QAM all-optical OFDM systems are investigated and compared. The phase noise compensation method is presented. Performance of All-optical OFDM signal before and after compensation signal for mitigating nonlinear phase noise is investigated. Numerical results of phase noise variation showed increasing the length from 200 km to 1200 km lead the phase noise variation from 0.001 to 0.03 rad2. It was shown that by increasing the fiber length the noise has higher variance in terms of phase than amplitude. Power effect and channel number were investigated in all-optical OFDM transmission system.

Keywords: 4-QAM, 16-QAM, phase noise estimation, phase noise compensation

4.1. PHASE NOISE EFFECTS IN ALL-OPTICAL OFDM

The above is the general picture for wave propagation through the fiber, but the aim of these scientific studies is to survey the effect of phase noise for All-optical OFDM, then propagation of each subcarrier will need to consider in previously mentioned equations. The OFDM based band signal s (t) is equal to M, is the number of sampling [1, 2],

$$u(0,t)=s(t)=\sum_{m=0}^{M-1}\sum_{i=0}^{N_{sc}-1} s_i(t_m) \quad , \quad s_i(t_m)=u_i(z\ ,t_m) \quad , \quad i=0,1,\dots,$$

$$hN_{sc} - 1 \tag{4.1}$$

By implementing the Cooley-turkey IFFT algorithm with a sample rate of T_s/N_{sc} [1], which T_s is OFDM symbol period, the mth sample of s (t) of one OFDM symbol readily yielding

$$s_m =$$

$$\begin{cases} 1/\sqrt{2}\left\{\frac{1}{\sqrt{N}}\sum_{n=0}^{N_{sc}/2-1} S_{2n}\ e^{\frac{j2\pi mn}{N_{sc}}} + \frac{1}{\sqrt{N}}\sum_{n=0}^{N_{sc}/2-1} S_{2n+1}e^{\frac{j2\pi m(2n+1)}{N_{sc}}}\right\} m \in [0, N_{sc}/2] \\ 1/\sqrt{2}\left\{\frac{1}{\sqrt{N}}\sum_{n=hN_{sc}/2}^{N_{sc}-1} S_{2n}\ e^{\frac{j2\pi mn}{N_{sc}}} + \frac{1}{\sqrt{N}}\sum_{n=hN_{sc}/2}^{N_{sc}-1} S_{2n+1}e^{\frac{j2\pi m(2n+1)}{N_{sc}}}\right\} m \in [N_{sc}/2, N_{sc} - 1] \end{cases}$$

$$\tag{4.2}$$

For the single span all-optical OFDM transmission system with omitting FWM effect the Eq. (4.2) can be rewritten as

$$u_i(z,t)=u_i(0,t)\ exp(j\phi_{iNL}) = s_i(t)\ exp(j\phi_{iNL}) \tag{4.3}$$

Total nonlinear phase shift for the OFDM can be extracted from the Eq. (4.2) by assuming that initial phase is equal to zero ($\phi_0 = 0$) and single span system as [3]

$$\Delta\phi_{NL} = \sum_{i=0}^{hN_{sc}-1} \phi_{iNL} = \frac{3}{8}\frac{\chi^{(3)}\omega_0}{k_0c^2}\left\{\frac{2N_{sc}-1}{N_{sc}}\sum_{i=0}^{N_{sc}-1}|u_i(0,t)|^2\ Z_{eff}\right\} \tag{4.4}$$

As can be concluded from this Eq. (4.4), $\Delta\phi_{NL}$ is time dependence, which leads to phase-induced spectral widening. In fact, temporally time varying phase brings about instantaneous optical frequency differs the key and trailing border from its main frequency ω_0. In other words, the central perhaps the OFDM signal will certainly acquire more phases rapidly versus across part.

4.2. PHASE NOISE ESTIMATION AND COMPENSATION FOR OPTICAL QAM SYSTEMS

The actual phase error cause of phase offset, frequency offset as well as laser phase noise needs to be compensated for the in-phase as well as quadrature information. After sampling from the received in-phase I*(t) as well as quadrature photocurrents Q*(t) defined by:

$$I^*(t) =$$
$$R.\left\{\sqrt{\frac{p_s}{2}}.I(t).\sqrt{P_{LO}}.\cos(\Delta\emptyset) + \sqrt{\frac{p_s}{2}}.Q(t).\sqrt{P_{LO}}.\sin(\Delta\emptyset)\right\} + n_1(t) -$$
$$n_3(t); \tag{4.5}$$

$$Q^*(t) = R.\left\{-\sqrt{\frac{p_s}{2}}.I(t).\sqrt{P_{LO}}.\sin(\Delta\emptyset) + \right.$$
$$\left.\sqrt{\frac{p_s}{2}}.Q(t).\sqrt{P_{LO}}.\cos(\Delta\emptyset)\right\} + n_2(t) - n_4(t); \tag{4.6}$$

So, we have the complex samples from the received complex cover as

$$A_K^* = I_K^* + jQ_K^* = C.(I_K + jQ_k).e^{-j\Delta\emptyset_k} + n_k^{tot} =$$
$$C.(a_k.e^{j\varphi_k}).e^{-j\Delta\emptyset_k} + n_k^{tot} = a_k^*.e^{j\varphi_k} \tag{4.7}$$

where C is a constant and n_k^{tot} demonstrates the complex shot noise that defined by

$$C = R.\sqrt{\frac{P_s}{2}} \cdot \sqrt{P_{LO}} , n_k^{tot} = (n_{1,k} - n_{3,k}) + j(n_{2,k} - n_{4,k}) =$$
$$n_{13,k} + jn_{24,k} \tag{4.8}$$

The actual amplitude and phase from the received complex envelope could be calculated using (4.5) and (4.6) as

$$a_k^* = \sqrt{(I_k^*)^2 + (Q_k^*)^2}$$
$$= \sqrt{\left(C.I_k.\cos(\Delta\emptyset_k) + C.Q_k.\sin(\Delta\emptyset_k) + n_{13,k}\right)^2 + \left(-C.I_k.\sin(\Delta\emptyset_k) + C.Q_k.\cos(\Delta\emptyset_k) + n_{24,k}\right)^2}$$

$$\tag{4.9}$$

$$\varphi_k^* = \tan^{-1}\frac{Q_k^*}{I_k^*} = \tan^{-1}\frac{-C.I_k.\sin(\Delta\emptyset_k) + C.Q_k.\cos(\Delta\emptyset_k) + n_{24,k}}{C.I_k.\cos(\Delta\emptyset_k) + C.Q_k.\sin(\Delta\emptyset_k) + n_{13,k}} \tag{4.10}$$

The total phase error could be generally defined since the difference between the actual received phase, φ_k^* and also the instant modulation position, φ_k.

$$\Delta\varphi_c^{tot} = \varphi_c^o - \varphi_c \tag{4.11}$$

The actual simulations had been carried out using VPI transmission maker. The MATLAB software was used to calculate the total phase error.

4.3. RESULTS AND DISCUSSION

Lists of the parameters used for simulations shown in Table 4.1.

Table 4.1. Lists of the parameters of the simulation

γ	Nonlinear coefficient	1.2W^{-1}/Km
$\Delta\nu_{opt}$	Optical filter bandwidth	42.7 GHz
λ	Wavelength	1.55 um
α	Attenuation coefficient	0.25 dB/km
η_{sp}	Spontaneous emission factor	1.41

Attenuation of the fiber is completely compensated by the EDFAs and the ASE noise added to the system is given by

$$\sigma^2 = 2S_{sp}\Delta\nu_{opt} = 2h\nu\eta_{sp}\Delta\nu_{opt}\alpha L_A \qquad (4.12)$$

The gain of each amplifier is equal to αL where L is fiber length between each amplifier. Lumped amplification is implemented in our long-haul fiber optical communication links. Each amplifier is placed after 80 kilometers' fiber length in our designed All-Optical OFDM communication systems. The gain of each amplifier is equal the loss in the span between the EDFAs. The channels spacing of $\Delta f = 25$ GHz is set in all of simulation results.

4.3.1. Fiber Optic Length Effects

Figure 4.1 shows the 4 QAM constellation diagram of all-optical OFDM versus increasing system lengths. The fiber length set from 160 km to 1120 km with the step of fiber spam length of 80 km. The comb generator laser launched powers is set at 10 mW (10 dBm). As it is apparent, due to soliton effects, the positive chirp effect from Phase noise compensates by positive chirp effect from the dispersion. To measure the real phase noise in the system, fiber dispersion coefficient set at zero. By increasing the fiber length, the noise has high variance in terms of phase. These phenomena effects on bit error rate of the system. Based on 4.2 sections the phase noise for optical QAM estimated and compensated. The phase noise compensated signal for relevant lengths are depicted in Figure 4.2.

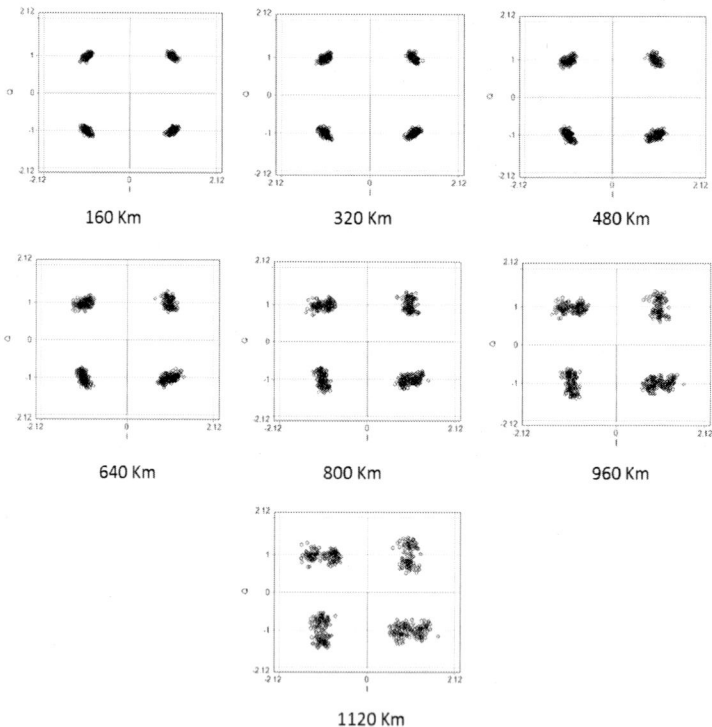

Figure 4.1. 4 QAM constellation diagram of all-optical OFDM vs. system lengths.

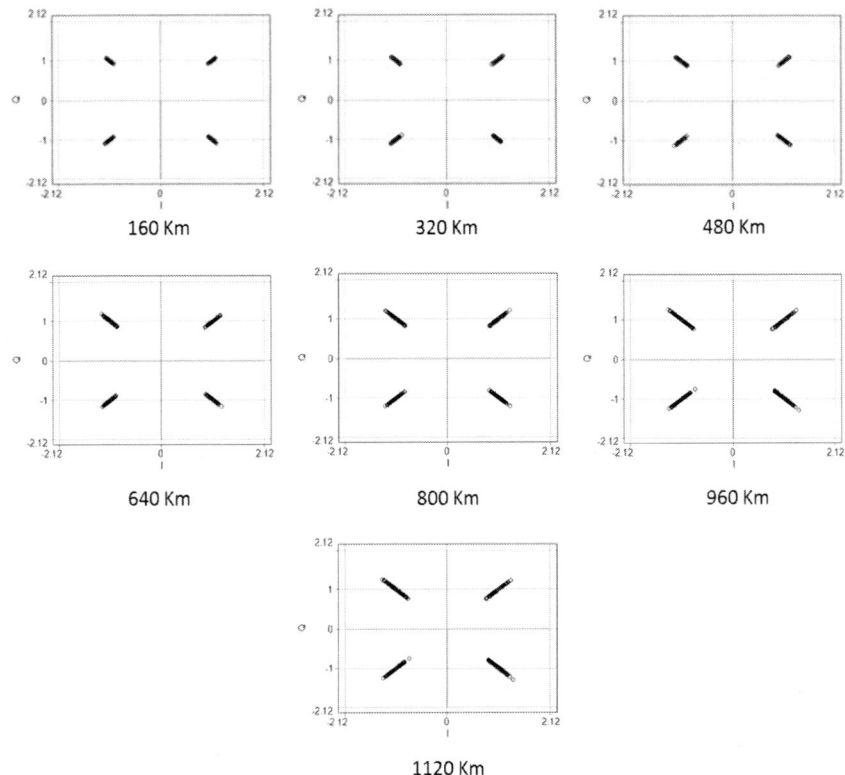

Figure 4.2. The phase noise compensated signal for relevant lengths

Figure 4.3 compares the performance of All-optical OFDM signal before and after compensation signal for mitigating nonlinear phase noise. The transmission distance is from 160 km to 1120 km and with all constellations normalized at the same transmit power. With square 4 QAM constellations, Bit error rate (BER) start around 10^{-20} for the length of 160 km and at length of 1120 km reached to 10^{-4}. With nonlinear phase compensation technique, the phase noise decrease to 10^{-50} at 160 km and at length of 1120 km reach to 10^{-4}. As can be seen, the phase noise compensation technique do no not effective at high BER.

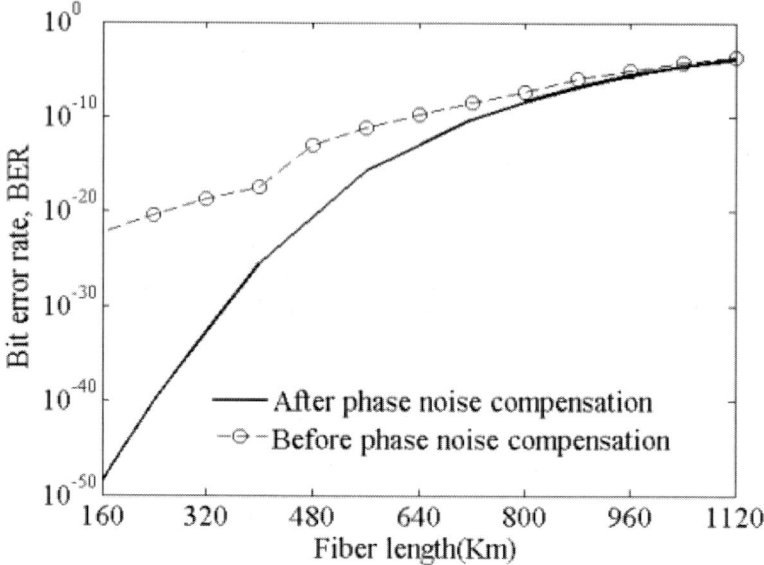

Figure 4.3. BER of All-optical OFDM signal before and after phase noise compensation.

4.3.2. Error Vector Magnitude (EVM)

Error Vector Magnitude (EVM) is a measure used to quantify the quality or performance of a modulated signal from a transmitter or receiver. EVM is expressed as the root mean square (RMS) value of the difference between a Received symbols and ideal symbols. As can be seen in Figure 4.4 EVM can be defining numerically as [4]:

$$EVM_{RMS} = \frac{\frac{1}{N}\sum_{r=1}^{N}|s_{ideal,r}-s_{meas,r}|^2}{\frac{1}{N}\sum_{r=1}^{N}|s_{ideal,r}|^2} \tag{4.13}$$

where $s_{meas,r}$ is normalized r^{th} symbol in a stream of measured symbols, $s_{ideal,r}$ is the ideal normalized constellation point for the rth symbol, and N is the number of a unique symbol in the constellation. Figure 4.5 shows the changes of EVM versus fiber length before and after compensation signal for mitigating nonlinear phase noise. As expected from consolation diagram, the (EVM) Increase as the fiber length increase.

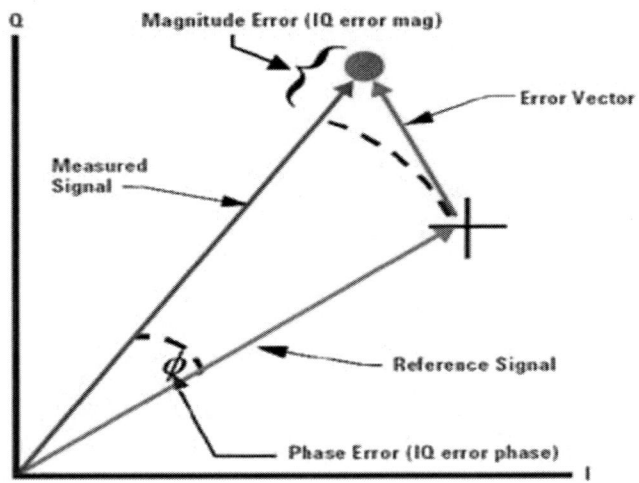

Figure 4.4. Error vector magnitude (EVM) diagram.

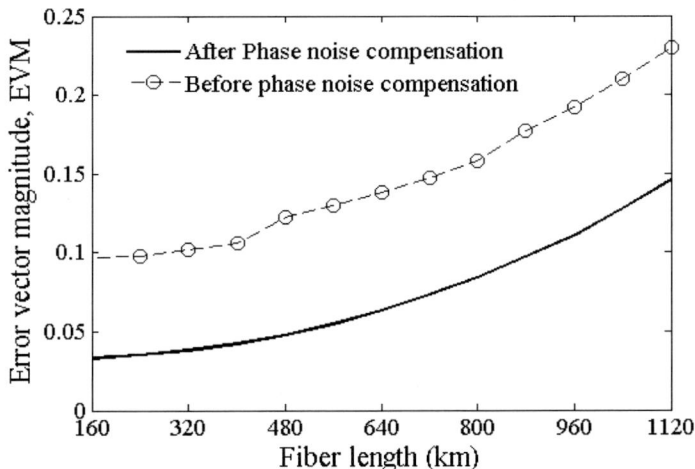

Figure 4.5. The changes of EVM vs. fiber length before and after phase noise compensation.

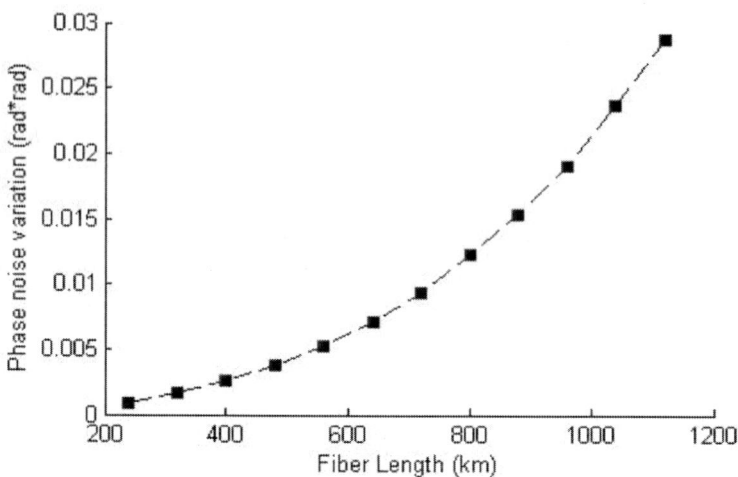

Figure 4.6. Phase noise variation vs. fiber length.

Increasing the phase noise at higher transmission length can be explained by phase noise variance versus fiber length. Figure 4.6 depicted the numerical results of phase noise variation versus fiber length. Increasing the length from 200 km to 1200 km lead the phase noise variation from 0.001 to 0.03 rad^2. These phenomena affect the error vector magnitude and increase the bit error rate of the system at higher lengths.

Figure 4.7 shows the 16 QAM constellation diagram of all-optical OFDM versus increasing system lengths. The fiber length set from 80 km to 560 km with the step of fiber span length of 80 km. The channels spacing of $\Delta f = 25$ GHz is set in all of simulation results. The phase noise compensated signal for the same lengths are depicted in Figure 4.8. The comb generator laser launched powers is set at 10 mW. As it is apparent, by increasing the fiber length the noise has higher variance in terms of phase than amplitude. These phenomena effects in bit error rate of system.

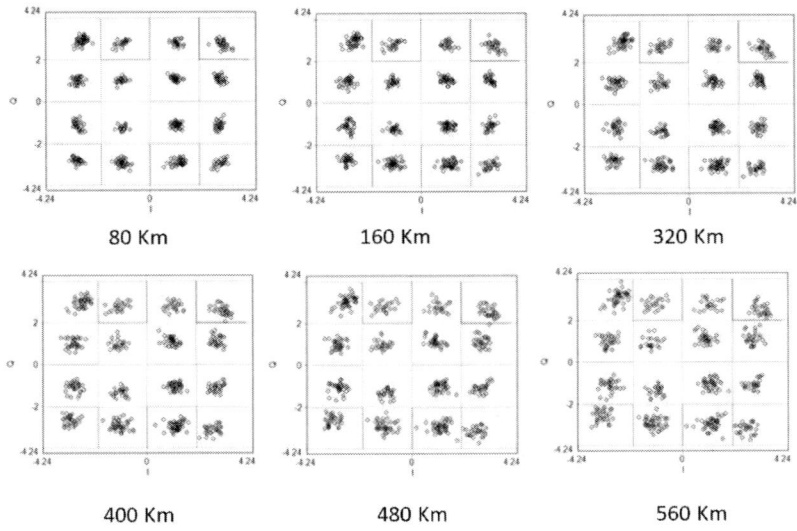

Figure 4.7. 16 QAM constellation diagram of all-optical OFDM vs. system lengths.

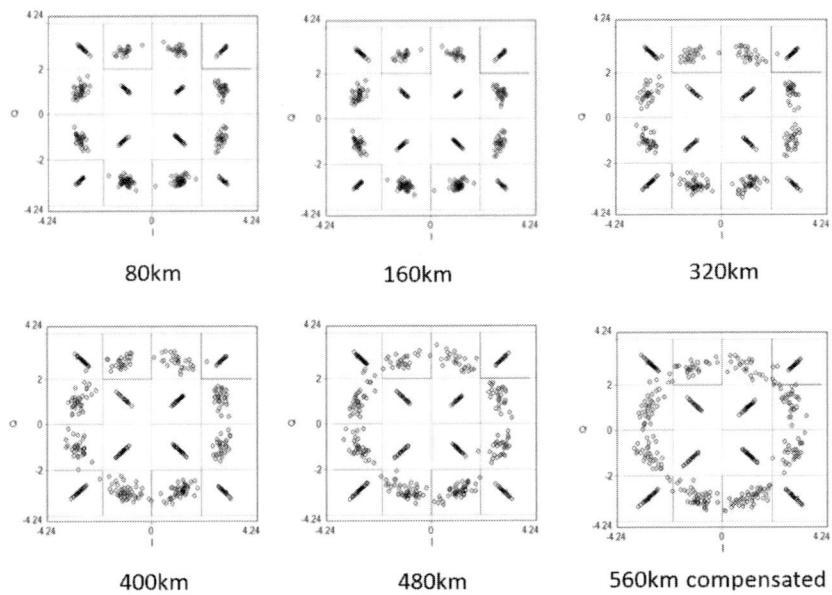

Figure 4.8. The phase noise compensated signal vs. system lengths.

Figures 4.9 and 4.10 show the variation of BER and EVM versus fiber length before and after phase noise compensation technique respectively. With the same explanation for 4 QAM and as expected from consolation diagram, the EVM and BER Increase as the fiber length increase.

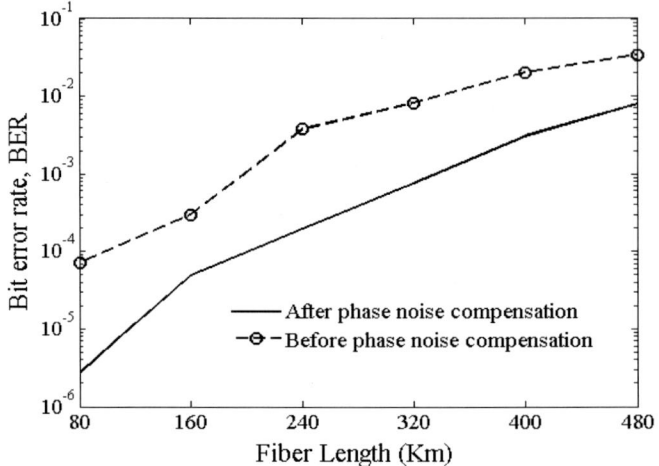

Figure 4.9. BER vs. fiber length before and after phase noise compensation.

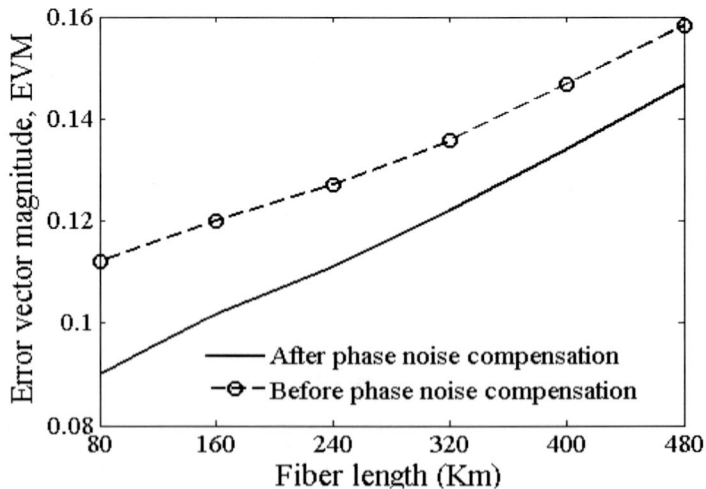

Figure 4.10. EVM vs. fiber length before and after phase noise compensation.

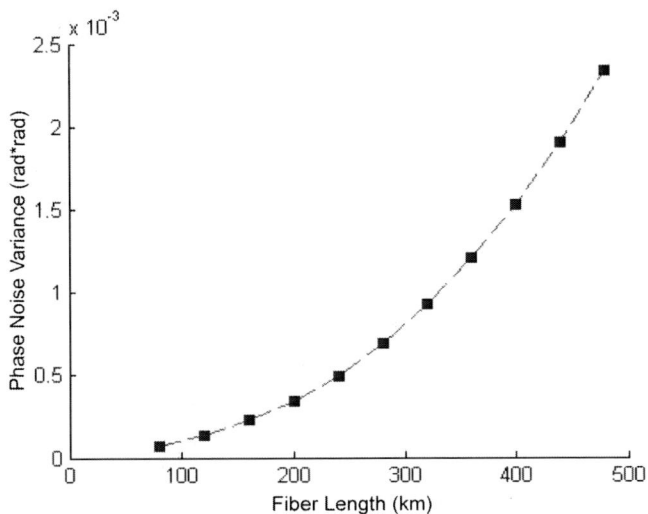

Figure 4.11. Phase noise variation of 16 QAM all-optical OFDM vs. fiber length.

Figure 4.11 depicted the numerical results of phase noise variation of 16 QAM all-optical OFDM versus fiber length. Increasing the length from 80 km to 480 km lead the phase noise variation from 0.001 to 0.0025 rad². These phenomena affect the error vector magnitude and increase the bit error rate of the system at higher lengths.

4.3.3. Effects of Channel Numbers and Laser Power

In the previous section, we discussed how phase noise affects propagated OFDM pulse through the medium. In this section, we investigate the effect of power and channel number in all-optical OFDM transmission system. It is well known that higher number of channels lead more phase noise in term of XPM and FWM nonlinearities. On the other hand, signal power is the main factors in nonlinear fiber optics. Therefore, there is more phase noise distortion at higher signal power for a higher number of the channel rather than the lower number of channels. To prove this possibility, the simulation results of all-optical OFDM with 3, 7 and 29 channel number in term of laser pump power variation from 10 mW to 40 mW is demonstrated in Figures 4.12-14.

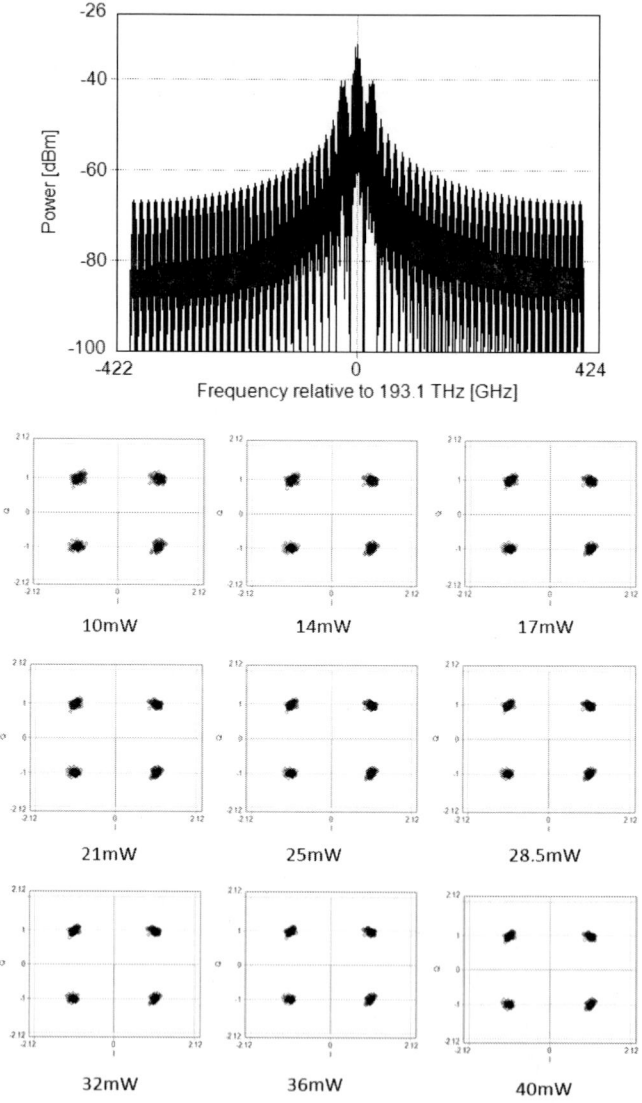

Figure 4.12. Channel spectrum and the constellation diagrams of 3 Channels all-optical OFDM vs. system power.

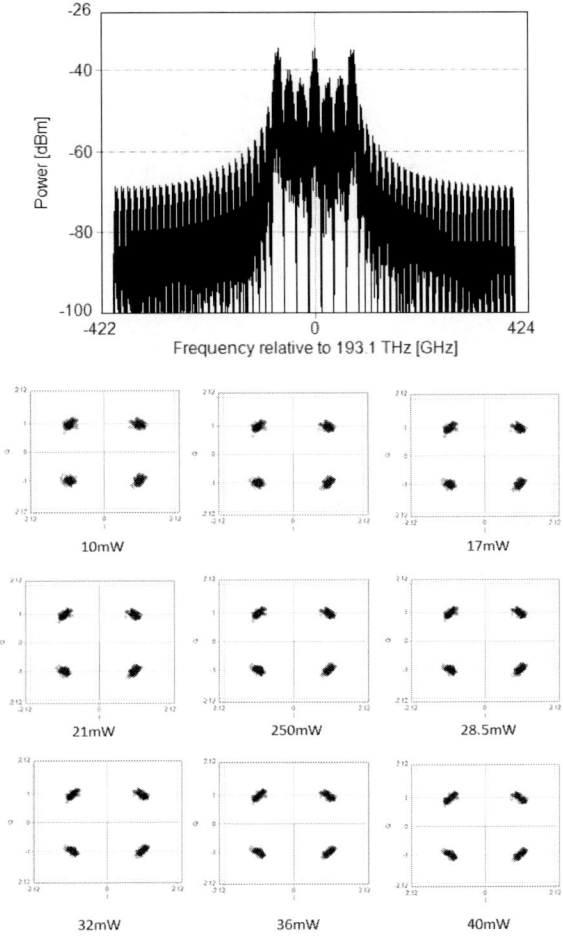

Figure 4.13. Channel spectrum and the constellation diagrams of 7 Channels all-optical OFDM vs. system power.

Figure 4.15 shows the detected constellation diagrams after Electro-Absorption Modulator (EAM) with respect to laser input power when the 4 QAM OFDM base-band signal. As it is clear from the Figures 4.15 and 4.16 with the increasing laser input power; there is a minor change in phase variation. This is shown that in this situation the XPM and SPM are less phase noise effects due to a low number of channels. The compensated signal cancellation diagram is depicted in Figures 4.15.

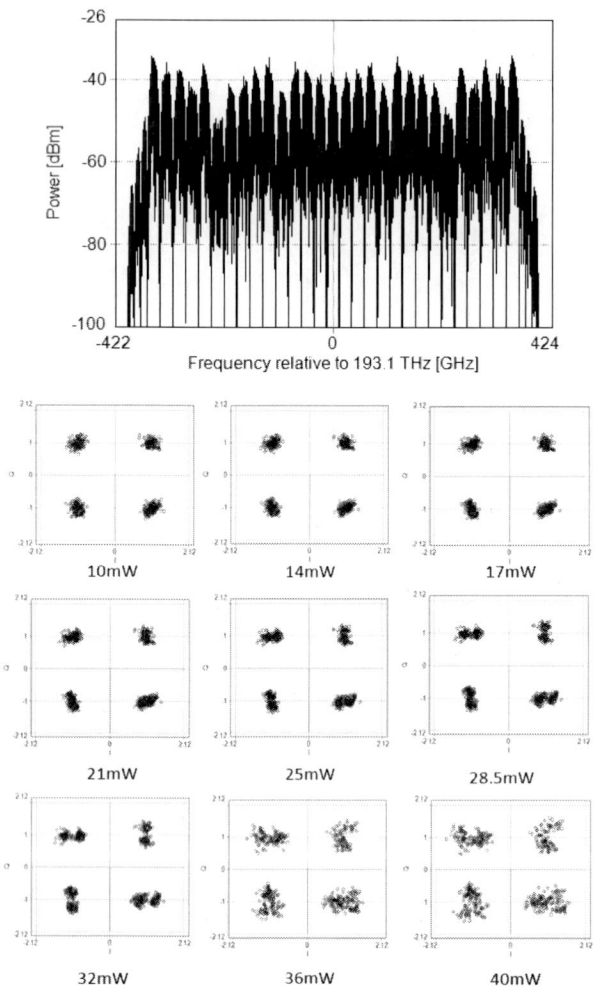

Figure 4.14. Channel spectrum and the constellation diagrams of 29 Channels all-optical OFDM vs. system power.

Figure 4.16 shows the bit error rate of 4 QAM all-optical OFDM with a channel number of 3. Interestingly by increasing the signal power the bit error rate decrease. This phenomenon is results of a higher power of the signal at receiver and improvement of photodetector sensitivity.

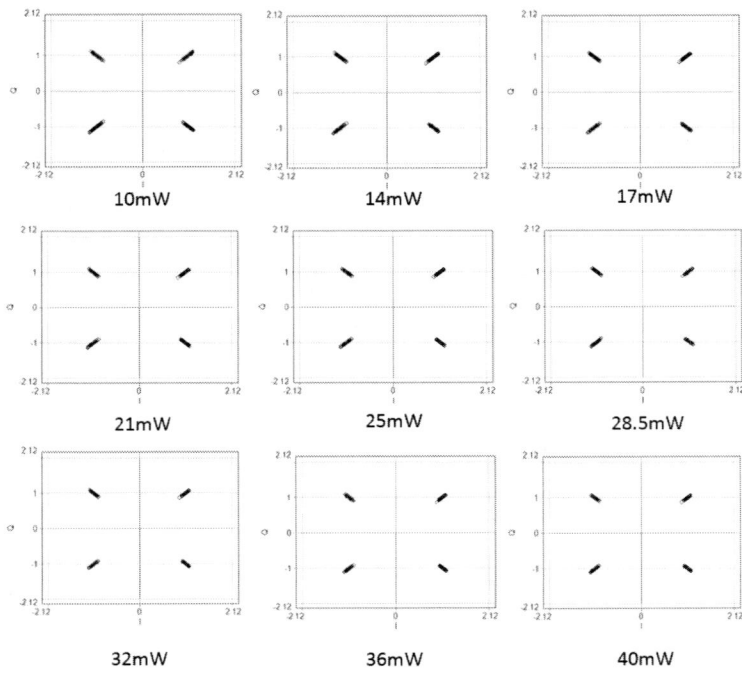

Figure 4.15. Phase noise compensated signal of 3 Channels all-optical OFDM vs. laser input power.

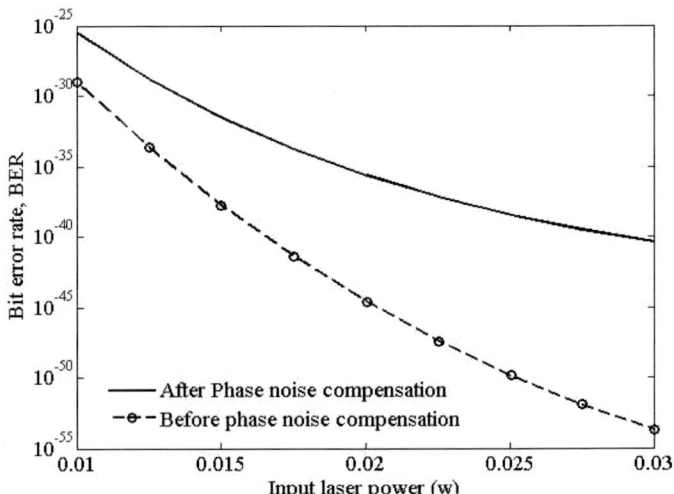

Figure 4.16. BER of 4 QAM all-optical OFDM with 3 numbers of channels.

To present the effect of the channel in phase noise effects the results of all-optical OFDM with 7 channels is presented in this section. The optical spectrum is shown in Figure 4.13. Figures 4.13, show the detected constellation diagrams before phase noise compensation and Figures 4.17, depicts them after phase noise compensation with respect to laser input power when the 4 QAM OFDM base-band signal. As can be clear from the Figures 4.17 and 4.18 with the increasing laser input power, there are changes in phase variation become obvious. This is shown that after 17mW the effects of phase noise due to XPM and SPM is dominant.

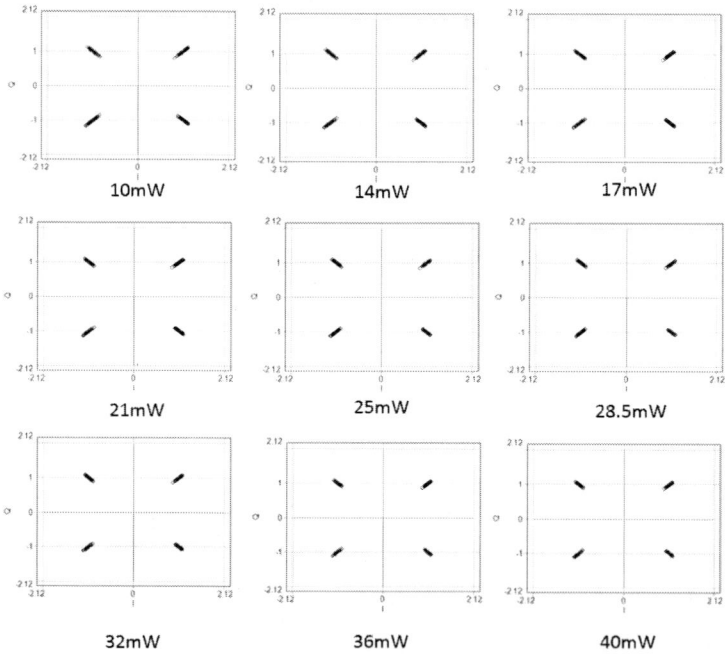

Figure 4.17. Phase noise compensated signal of 7 Channels all-optical OFDM vs. laser input power.

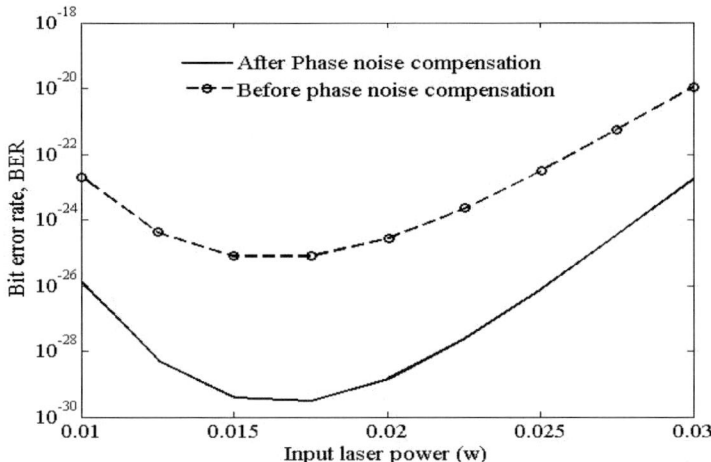

Figure 4.18. BER of 4 QAM all-optical OFDM with 7 numbers of channels.

Figure 4.18 shows the bit error rate of 4 QAM all-optical OFDM with a channel number of 7. By increasing the signal power from 0.01 w to 0.0175 w, the bit error rate decreases because of a higher power of the signal at receiver and increase of the system performance. Above the 0.0175 w laser power, the phase noise dominates. This phenomenon leads BER rate increasing at higher power.

At the end, the results of maximum 29 channel number are presented in this section. The optical spectrum is shown in Figure 4.14. Figure 4.14 shows the detected constellation diagrams before phase noise compensation and Figure 4.19 shows them after phase noise compensation with respect to laser input power when the 4 QAM OFDM base-band signal. As can be clear from the Figures 4.19 and 4.20, with the increasing laser input power, phase variation increased as expected. This is shown that the effects of phase noise due to FWM are dominant in all power ranges.

Figure 4.20 shows the bit error rate of 4 QAM all-optical OFDM with 29 channel number. By increasing the signal power from 0.01 w to 0.04 the BER rate increasing.

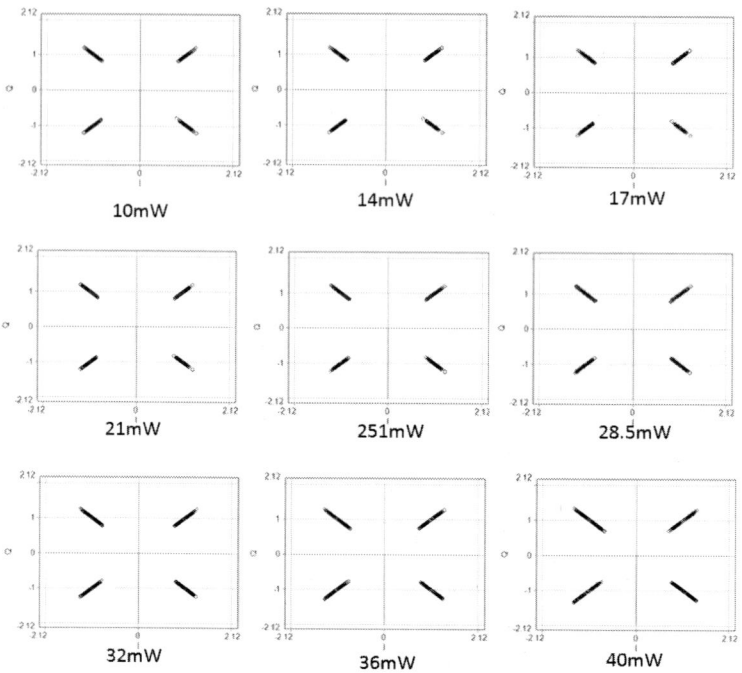

Figure 4.19. Phase noise compensated signal of 29 Channels all-optical OFDM vs. laser input power.

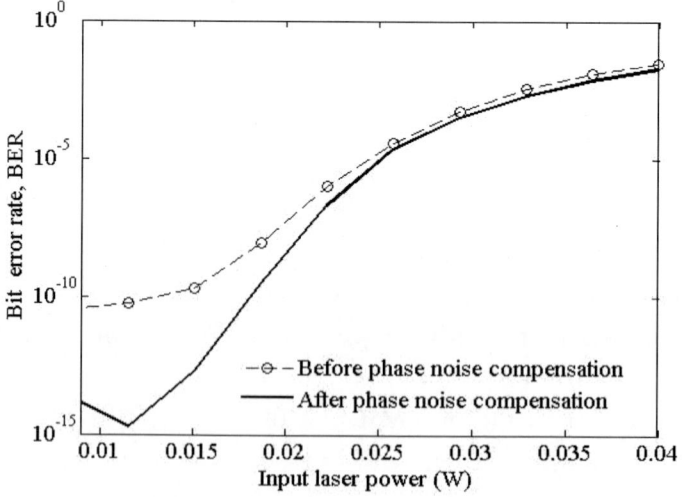

Figure 4.20. BER of 4 QAM all-optical OFDM with 29 numbers of channels.

In conclusion, the effect of phase noise on both the 4-QAM and 16-QAM all-optical OFDM systems are investigated and compared. The phase noise compensation method is presented. Performance of All-optical OFDM signal before and after compensation signal for mitigating nonlinear phase noise is investigated. With square 4 QAM constellation, at transmission distance from 160 km to 1120 km Bit error rate (BER) changed from 10^{-20} to 10^{-4}. With nonlinear phase compensation technique, the phase noise decrease to 10^{-50} at 160km and at length of 1120 km reach to 10^{-4}. Increasing the phase noise at higher transmission length was explained by phase noise variance. Numerical results of phase noise variation showed increasing the length from 200 km to 1200 km lead the phase noise variation from 0.001 to 0.03 rad^2. The 16 QAM constellation diagram of all-optical OFDM was analyzed. The comb generator laser launched powers was set at 10 mW. It was shown that by increasing the fiber length the noise has higher variance in terms of phase than amplitude. These phenomena effects on bit error rate of the system. Increasing the length from 80 km to 480 km leads the phase noise variation from 0.001 to 0.0025 rad^2. This phenomenon affects the error vector magnitude and increases the bit error rate of the system at higher lengths. Effect power and channel number were investigated in all-optical OFDM transmission system. It is shown that higher number of channels lead more phase noise in term of SPM, XPM and FWM nonlinearities. On the other hand, signal power was the main factors in nonlinear fiber optics. Therefore, there is more phase noise distortion at higher signal power for a higher number of the channel rather than the lower number of channels.

REFERENCES

[1] H. Chen, M. Chen & S. Xie, (2009) All-optical sampling orthogonal frequency-division multiplexing scheme for high-

speed transmission system, *Lightwave Technology, Journal of*, 27(21), 4848-4854.

[2]   S. Kumar, *Impact of Nonlinearities on Fiber Optic Communications* vol. 7: Springer Science & Business Media, 2011.

[3]   M. Nazarathy, J. Khurgin, R. Weidenfeld, Y. Meiman, P. Cho, R. Noe, I. Shpantzer & V. Karagodsky, (2008) "Phased-array cancellation of nonlinear FWM in coherent OFDM dispersive multi-span links," *Optics express*, 16(20), 15777-15810.

[4]   G. P. Agrawal, *Fiber-optic communication systems* vol. 1, 1997.

ABOUT THE AUTHORS

Dr. Iraj Sadegh Amiri
Assistant Professor
Computational Optics Research Group
and
Faculty of Applied Sciences
Ton Duc Thang University
Ho Chi Minh City, Vietnam
Tel: +16179599872 / Email: irajsadeghamiri@tdt.edu.vn

Dr. Amin Khodaei
Lecturer
Photonics Research Centre
Faculty of Science
University of Malaya, Kuala Lumpur, Malaysia

Dr. Volker J Sorger
Department of Electrical and Computer Engineering
The George Washington University
Washington, D.C., US

INDEX